Review
k12.princetonreview.com

Know It All!
Grades 3–5 Math
by Lisa Meltzer

Random House, Inc.
New York

www.randomhouse.com/princetonreview

nation's leaders in test preparation. The Princeton Review helps millions of students every year prepare for standardized assessments of all kinds. The Princeton Review offers the best way to help students excel on standardized tests.

The Princeton Review is not affiliated with Princeton University or Educational Testing Service.

Princeton Review Publishing, L.L.C.
160 Varick Street, 12th Floor
New York, NY 10013

E-mail: textbook@review.com

Published in the United States by Random House, Inc., New York.

ISBN 0-375-76375-9

Editor: Linda Fan
Development Editor: Scott Bridi
Production Editor: Lisbeth Dyer
AVP and Associate Publisher: Iam Williams
Art Director: Neil McMahon
Director of Creative Services: Mike Rockwitz
Production Coordinator: Carmine Raspaolo

Manufactured in the United States of America

9 8 7 6 5 4 3 2 1

First Edition

Contents

Introduction for Parents and Teachers

Introduction for Parents and Teachers

About This Book

Know It All! focuses on essential math skills that students need in order to succeed in school and on standardized achievement tests while providing information about a wide array of subjects. The math skills that students will review and practice in this book are based on the math standards developed by the National Council of Teachers of Mathematics.

Know It All! contains chapters covering essential math skills, reviews called Brain Boosters, an answer key for the chapters and Brain Boosters, a practice test, and answers and explanations for the practice test. Each **chapter** focuses on a skill or set of related skills, such as the chapter about whole numbers or the chapter about operations and properties of numbers. Each **Brain Booster** includes practice questions that review the content in the chapters that precede it. The **answer key** provides correct answers to the questions in the chapters and Brain Boosters. The **practice test** is similar to actual standardized achievement tests in style, structure, difficulty level, and skills tested. The **answers and explanations** provide correct answers to the questions on the practice test and explain how students can answer them correctly.

Each **chapter** contains the following:

- an introduction that presents the content covered in the chapter and definitions of relevant terms
- a sample passage and question followed by a step-by-step explanation of how to answer the question
- passages about interesting subjects and practice questions that cover the content of the chapter

Some chapters include *Know It All!* tips to assist students in further developing their skills. There will also be cumulative review sections called Brain Boosters following every few chapters in the book.

The **practice test** contains the following:

- multiple-choice, short-answer, and open-response questions that are similar to questions on actual standardized achievement tests
- a bubble sheet similar to bubble sheets on actual standardized achievement tests on which students can fill in their answers to multiple-choice questions

Answers and explanations following the practice test illustrate the best methods to solve each question.

About The Princeton Review

The Princeton Review is one of the nation's leaders in test preparation. We prepare more than two million students every year with our courses, books, on-line services, and software programs. We help students around the country on many statewide and national standardized tests in a variety of subjects and grade levels. Additionally, we help students on college entrance exams such as the SAT-I, SAT-II, and ACT. Our strategies and techniques are unique and, more importantly, successful. Our goal is to reinforce skills that students have been taught in the classroom and show them how to apply these skills to the format and structure of standardized tests.

About Standardized Achievement Tests

Across the nation, different standardized achievement tests are used in different locations to assess students. States choose what tests they want to administer, and often, districts within the state choose to administer additional tests. Some states administer state-specific tests, which are given only in that state and linked to that state's curriculum. Examples of state-specific tests are the Florida Comprehensive Assessment Test (FCAT) and the Massachusetts Comprehensive Assessment System (MCAS). Other states administer national tests, which are used in several states in the nation. Examples of some national tests are the Stanford Achievement Test (SAT9), Iowa Test of Basic Skills (ITBS), and TerraNova/CTBS (Comprehensive Test of Basic Skills). Some states administer both state-specific tests and national tests.

Most tests administered to students contain multiple-choice and open-response questions. Some tests are timed; others are not. Some tests are used to determine whether a student can be promoted to the next grade; others are not. None of the tests can assess all of the unique qualities of your students. They are intended to show how well students can apply skills they have learned in school in a testing situation.

State-specific tests, because they are connected to a state curriculum, show how well students can apply the curriculum that was taught in their school in a testing situation. National tests are not connected to specific state's curricula but have been created to include content that most likely would be taught in your student's grade and subject. Therefore, sometimes a national test administered to your student will test content that has not been taught in your student's grade or school. National tests show how well a student has done on the test in comparison with how well other students in the nation have done on that test.

To find out what test(s) your student will take, when the test(s) will be given, whether a test is timed, whether it affects grade promotion, or other information, contact your school or your local school district. To find out more about state-specific tests, go to www.nclb.gov/next/where/statecontacts.html. You can also click on "Assessment Advisor" at the Web site www.k12.princetonreview.com.

Student
Introduction

Student Introduction

About This Book

What kind of person is a *know it all?* Someone who is hungry for more information and wants to learn new things is a *know it all.* Someone who is amazed by what he or she learns is a *know it all.* Who can be a *know it all?* Anybody—including you!

This book, *Know It All!*, is an adventure for your mind. *Know It All!* is chock-full of wild, weird, fascinating, and unbelievable articles—all of which contain true information! Plan to be surprised, amused, and grossed out on this adventure.

In addition to feeding your brain all sorts of interesting information, *Know It All!* teaches test-taking tips and gives you test practice. By the end of this book, you will have the biggest, strongest brain ever! You'll be ready for the Brain Olympics. You'll be a *know it all!*

Know It All! contains **chapters, brain boosters,** and a **practice test.**

Each **chapter**
- defines a skill or group of skills, such as the chapter about whole numbers
- shows you how to use the skills to answer a sample question
- provides passages and practice questions like those you may see on standardized tests in school
- may have *Know It All!* tips to help you become a *know it all*

Each **brain booster**
- reviews skills from the preceding chapters
- includes fun and interesting passages to read
- provides practice questions for you to answer

The **practice test**
- gives you practice answering questions similar to those on standardized tests
- provides a bubble sheet similar to one you'll see on standardized tests

This book has an **answer key** for the questions in the chapters and Brain Boosters. There is also an **answers and explanations** section that shows you correct answers to the questions on the practice test and explains how you can answer them correctly.

About Standardized Achievement Tests

Standardized achievement tests. Who? What? Where? When? Why? How?

You know about them. You've probably taken them. But you might have a few questions about them. If you want to be a *know it all,* then it would be good for you to know about standardized achievement tests.

Standardized tests may be different from the tests that you usually take. These may be the only tests you take that are not written by your teachers. Everyone in your grade at your school will probably take the same test. In fact, everyone in your grade in your state might take the same test. Some of the standardized tests you take will be timed, and others may not be.

To find out about any standardized achievement test, ask your teachers and parents questions about it. The following list shows you some questions that you might want to ask:

Who?   You!

What?   What kinds of questions will be on the test?

              What kinds of skills will be tested by the test?

              What do I have to bring to the test?

Where?   Where will I take the test?

When?   When will I take the test?

Why?   Why am I taking the test?

How?   How should I prepare for the test?

              How much time will I have to finish the test?

No test can assess your individual qualities as a *know it all.* The purpose of a standardized achievement test is to show how well you know the skills that you learned in school.

About the Icons in This Book

This book contains many different small pictures, which are called icons. The icons tell you about the topics of the articles in the book.

 Alternative Animals Read these passages to learn about strange animals and their wild habits. You'll learn about the biggest, smallest, oldest, fastest, and most interesting animals on the planet. You can find these passages in chapters 3, 7, 14, and 15, and in Brain Boosters 1, 2, and 4.

 Hip History Learn about cool castles and surprising people from the past. These passages will teach you some of the most interesting stories in history. You can find these passages in chapters 3, 5, 6, 9, and 13.

 For Your Amusement You want to play games? Read these passages to learn about fun games, toys, amusement parks, and festivals. You can find these passages in chapters 3, 7, 9, 11, and 18.

 Extreme Sports Read these passages to learn about outrageous contests, wacky people, and incredible achievements in the world of sports. You may not have even heard of some of these sports! You can find these passages in chapters 5 and 14, and in Brain Boosters 2, 3, and 4.

 Grosser Than Gross How gross can you get? Read these passages if you want to learn about really gross things. Be warned: Some of the passages may be so gross that they're downright scary. You can find these passages in chapters 4, 13, 14, 15, and 16, and in Brain Booster 1.

 Mad Science You'll see science in these passages like you've never seen it before. You'll learn about all sorts of interesting science-related stuff. You can find these passages in chapters 1, 8, 11, 16, and 17, and in Brain Booster 3.

Outer Space Oddities Do you ever wonder what goes on in the universe away from planet Earth? Satisfy your curiosity by reading these passages about astronomical outer space oddities. You can find these passages in chapters 9, 10, 13, and 17.

 Explorers and Adventurers Did you ever want to take a journey to learn more about a place? Well, you'll get the opportunity to do that when you read these passages about explorers and adventurers. You can find these passages in chapters 2, 7, and 12.

 The Entertainment Center Do you enjoy listening to music or watching television and movies? Well, here's your chance to read about them! You can find these passages in chapters 1, 6, 16, and 18.

 Art-rageous Are you feeling a bit creative? Read these passages to get an unusual look into art that's all around you in books, drawings, paintings, and much more. You can find these passages in chapters 2, 6, 8, and 10.

 Bizarre Human Feats People do very strange things. You can read about some incredible-but-true human feats in these passages. You can find these passages in chapters 4, 8, 11, and 15.

 Wild Cards You'll never know what you're going to get with these passages. It's a mixed bag. Anything goes! You can find these passages in chapters 1, 2, 4, 5, 10, 12, 17, and 18.

Things to Remember When Preparing for Tests

There are lots of things you can do to prepare for standardized achievement tests. Here are a few examples.

- **Work hard in school all year.** Paying attention in class and learning the skills that will be on the tests is a great way to prepare.

- **Read.** Read everything you can. Read your homework, your textbooks, the newspaper, magazines, novels, plays, poems, comics—even the back of the cereal box. Reading a lot is a great way to prepare for tests.

- **Ask your teachers and parents questions about your schoolwork whenever you need to.** Your teachers and your parents can help you with your schoolwork. Asking for help when you need it is a great way to prepare for tests and to become a *know it all*.

- **Work on this book!** This book provides you with loads of practice for tests. You've probably heard the saying "Practice makes perfect." Well, it's true! Practice can help you prepare for tests.

- **Ask your teachers and parents for information about the tests.** If you have questions about the tests, ask them! Being informed is a great way to prepare for tests.

- **Get enough sleep before you take a test.** Being awake and alert while taking tests is very important. Your body and your mind work best when you've had enough sleep. So get some rest on the nights before the tests!

- **Have a good dinner and a good breakfast before you take a test.** Eating well will fuel your body with energy, and your brain thrives on energy. You want to take each test with all the engines properly running in your brain.

- **Check your work.** When taking a test, you may end up with extra time. You could spend that extra time twiddling your thumbs. But if you use your extra time to check your work, you might spot some mistakes—and improve your score.

- **Stay focused.** You may find that your mind wanders away from the test once in a while. Don't worry—it happens. Just say to your brain, "Brain, it's great that you are so curious, imaginative, and energetic. But I need to focus on the test now." Your brain will thank you later.

CHAPTER 1
Question Types

How did raw eggs fly hundreds of feet into the air?

Just how big is a giant beach towel, anyway?

What made Adam Sandler decide to become a performer?

Do you want to do the best you can on any standardized test? Well, then you'll need to learn about the three basic types of math questions: multiple-choice, short-answer, and open-response questions.

Multiple-Choice Questions

Multiple-choice questions are questions that are followed by several answer choices. This type of question is pretty nifty. Why? Because the correct answer is right on the page! Your job is to figure out which answer choice is correct.

Here's the Know It All Approach to answering multiple-choice questions.

Know It All Approach

Step 1   **Read the question carefully.** This is the first step to answering any test question. If the question comes with a graph or a figure, you should look at it closely too.

Step 2   **Look for important words and numbers that you will need in order to answer the question.** Important words will tell you what you are trying to find or what operation to use. It may help you to notice words such as *not, each, total, least, greatest, approximately,* and *estimate.* Of course, you should also pay close attention to the numbers in a question.

Step 3   **Answer the question.** That sounds simple enough, right? When you answer the question, you should write your calculations clearly so that you can read what you've written. Then work carefully to figure out the answer.

Step 4   **Double-check your answer.** A good way to double-check your answer is to answer the question a second time. It's a good idea to reread the question and compare it to the information you used to figure out the answer. If you double-check your work, you will prevent mistakes.

Step 5   **Read all of the answer choices.** Even if you think you have the answer, it is best to read all of the answer choices in order to **be sure** that you have found the correct answer choice. Sometimes things just slip through the cracks. If you read each of the answer choices, you may notice something about a question that you hadn't noticed before.

Step 6   **For multiple-choice questions that you can't answer easily, just get rid of wrong answer choices.** Sometimes it is easier to find the correct answer choice if you get rid of answer choices that you *know* are wrong. Sometimes answer choices are way off the mark or just plain silly. If you see that an answer choice is wrong, you should get rid of it. If you can get rid of all of the answer choices except one, there's a pretty good chance that you've found the correct answer choice!

Step 7   **Fill in the bubble that goes with the answer choice you pick.** After you decide which answer choice is correct, fill in the bubble that goes with it.

Read the sample multiple-choice question below and see how you can use the Know It All Approach to answer it.

Egg Rockets

Teams of high school students recently took part in a contest to design, build, and launch their own rockets. The rules of the contest were specific. Students had to build a rocket from scratch. The rocket could not weigh more than 3.3 pounds. In addition, the rocket needed to carry two raw eggs exactly 1,500 feet into the air and then back to the ground without breaking them.

Directions: Answer the question based on the passage Egg Rockets.

▶   A group of students built three test rockets. The first rocket flew 1,456 feet into the air. The second rocket flew 1,549 feet. The third rocket flew 1,400 feet. Order the flight distances from least to greatest.

○ A 1,400; 1,549; 1,456
○ B 1,456; 1,549; 1,400
○ C 1,400; 1,456; 1,549
○ D 1,549; 1,456; 1,400

Know It All Approach

Step 1 **Read the question carefully.** This is the first thing you should do when you answer any test question. If you read carelessly, you might miss important words. That may cause you to select an incorrect answer choice. If the question includes a figure or a graph, pay close attention to that, too.

Step 2 **Look for important words and numbers you will need in order to answer the question.** The important numbers in this question are 1,456; 1,549; and 1,400. The important words are *order* and *least to greatest*. Most numbers in any math word problem will fall into the "important information" category.

Step 3 **Answer the question.** Take your time and write your work neatly. In this case, the question asks you to order the flight distances of the three rockets from least to greatest. To do this, write the important numbers from the question in a column. Make sure the place values of all the numbers are lined up like this.

1,456

1,549

1,400

Compare what you've written to the information in the question. Did you write everything correctly? If so, look at the thousands place (the one at the far left). Compare all of the numbers in that place value. They're all the same, so you still can't tell how to order the numbers.

Now move on to the hundreds place. The number 5 is greater than 4. So 1,549 is the greatest of the three numbers. It should be last in the list.

Now compare the tens place values of the other two numbers. The number 5 is greater than 0. This means that 1,456 is greater than 1,400 but less than 1,549. Now you have all the information you need to write your list. It should look like this.

1,400; 1,456; 1,549

Remember to write neatly. If you can't read your own handwriting, you could scribble numbers until your hand went numb and it wouldn't help you answer the question.

Step 4 **Double-check your answer.** Compare all of the numbers in the different place values to make sure you wrote the numbers in the correct order.

Step 5 **Read all of the answer choices.** Read **all** of the answer choices to **be sure** that you have found the correct one. In this case, compare your list—1,400; 1,456; 1,549—with each answer choice. Your list matches the order of the numbers in answer choice (C), so (C) is the correct choice!

Step 6 **If you just don't know which answer is correct, get rid of the wrong answer choices first.** Sometimes it's easier to find the correct answer choice if you get rid of the answer choices you **know** are wrong. Some questions include answer choices that are obviously way, *waaaaaaay* wrong. In this question, if you know that 1,400 is the first number, you can get rid of answer choices (B) and (D). Then you would be left with two answer choices to choose from

Step 7 **Fill in the bubble that goes with the answer choice you pick.** By now you're probably a whiz at coloring inside the lines, so this part is a breeze. After you decide which answer choice is correct, carefully fill in the bubble that goes with it.

Short-Answer Questions

Short-answer questions ask you to write an answer using your own words. Fortunately, the key word in *short-answer* is *short*, so you won't have to do much writing. Usually, you'll just have to write a number or a couple of words to answer these questions. You can handle that.

Here's the Know It All Approach to answering short-answer questions.

Step 1 **Read the question carefully.** This is *always* the first step when you answer any question on a test. If the question comes with a graph or a figure, you should look at that closely too.

Step 2 **Read the question again. Look for important words and numbers you will need in order to answer the question.** Important words will tell you what you are trying to find or what operation to use. It may help you to notice words such as *not, each, total, least, greatest, approximately,* and *estimate.* Of course, you should also pay close attention to the numbers in a question.

Step 3 **Answer the question.** This sounds like a good plan, right? Write out the information that you will need in order to answer the question, and then do the work to get your answer. Make sure that you've copied the information in the question correctly.

Step 4 **Double-check your answer.** It's a good idea to reread the question and compare it to the information you used to figure out the answer. If you double-check your work, you will prevent mistakes.

Step 5 **Write your final answer in the space provided.** Write the answer clearly and make sure you answer the question completely.

Read the sample short-answer question below and see how you can use the Know It All Approach to answer it.

One Big Beach Towel

On January 25, 2000, a giant beach towel was exhibited at a textile fair in Valencia, Spain. The towel was 9.4 meters wide and 14.46 meters long. The illustration below shows the measurements of the towel.

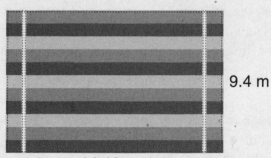

9.4 m

14.46 m

Directions: Answer the question based on the passage One Big Beach Towel.

▶ What was the perimeter of the beach towel?

Answer: _____

Know It All Approach

Step 1   **Read the question carefully.** In this case, the question also includes a diagram of the beach towel. Look at the information in the diagram closely, too.

Step 2   **Read the question again. Look for words and information that you will need in order to figure out the answer.** The important words and numbers in this problem are *9.4 meters*, *14.46 meters*, and *perimeter*.

Step 3   **Answer the question.** Here's the fun part. To find the perimeter of an object, you need to add the measurements of all of its sides. (It's okay if you didn't know that. Perimeter is covered in chapter 14.) Write down the addition problem and do the work neatly. Make sure that the numbers you've written match the ones in the problem and the diagram. Also make sure to line up any decimal points in the numbers.

$$
\begin{array}{r}
14.46 \\
14.46 \\
9.4 \\
+\ 9.4 \\
\hline
47.72
\end{array}
$$

Step 4   **Double-check your answer.** This will help you to avoid making any silly mistakes. Review your steps and answer the question a second time. Make sure you have included the correct lengths of all four sides of the beach towel in your addition problem. Then check your addition. Make sure you lined up the place values of all your numbers properly and that you remembered to regroup the numbers when necessary.

Step 5   **Write your final answer in the space provided.** Because this is a short-answer question, your answer is going to be—you guessed it—short! All you have to do now is write your final answer on the line provided. If you did your math correctly, you write "47.72 meters" on the line.

Open-Response Questions

Open-response questions ask you to write your answer and show your work. Sometimes these questions have more than one part for you to answer. It's important to look for questions that have more than one part. The steps to answering these types of questions are similar to the steps for short-answer questions. See the Know It All Approach on pages 15 and 16 for a review.

Read the sample open-response question below and see how the Know It All Approach can help you answer it.

Late Bloomer

 Many famous actors and actresses began dreaming of becoming performers when they were very young. Actor and comedian Adam Sandler was *not* one of these pint-size fame-seekers, however. Adam claims that he never even thought about becoming a comedian until he was 17 years old. Adam's brother was the one who first convinced him to perform at comedy clubs. The rest is history.

Directions: Answer the questions based on the passage Late Bloomer.

Part A

Josh, a big Adam Sandler fan, was born in 1996. In what year will Josh turn 18?

Show your work.

Answer: _____

Part B

On Josh's birthday in 2006, Adam Sandler will be 4 times as old as Josh. How old will Adam Sandler be?

Show your work.

Answer: _____

Know It All Approach

Step 1

Read the question carefully. You've heard this before—careful reading is your best friend when answering questions. It keeps you from overlooking information you may need in order to answer the question. Many open-response questions come with graphs or figures. Some open-response questions—like this one—also have more than one part for you to answer. Read the whole question carefully so you won't forget to answer any part of it.

Step 2

Read the question again. Notice the important words and information that you need in order to figure out the answer. First, you've probably noticed that this problem contains two parts: part A and part B. In part A, the important information is the year 1996 and the words *what year*, and *18th birthday*. In part B, note the years 2006 and 1996 and the words *4 times as old*.

Step 3

Use the space provided to figure out the answer. Write neatly and show all of your work. This is important. Always show your work for open-response questions. Even if you can't figure out the correct answer, you might get some credit for showing your work. Be sure to write neatly so the people who grade your test can read your work.

Part A asks you to figure out the year that Josh will turn 18. It tells you that Josh was born in 1996. That means you just have to add 18 years to 1996 to get your answer. $1996 + 18 = 2014$. So Josh will be 18 in the year 2014.

Part B asks you to figure out how old Adam Sandler will be on Josh's birthday in 2006. First, find out how old Josh will be in 2006. To do so, subtract the year Josh was born, 1996, from the year 2006. $2006 - 1996 = 10$. Josh will be 10 years old in the year 2006. The problem says that Adam Sandler will be 4 times as old as Josh in the year 2006. To figure out Adam Sandler's age in 2006, multiply Josh's age, 10, by 4. $10 \times 4 = 40$. That's your answer to part B. In the year 2006, Adam Sandler will be 40 years old.

Step 4

Double-check your answer. Review your steps and answer the question a second time. Make sure that the numbers you used to get your answer match the numbers in the word problem.

Step 5

Make sure you answer all parts of the question. Use your best handwriting to write the answer to the question. Make sure you answer both parts of the question. Also make sure that you write your answers neatly. The people who grade the test must be able to read your handwriting. If they can't read your writing, you won't get any points for your answer, even if it's correct.

Subject Review

In this chapter you learned about three different types of math questions. Review what you've learned so far.

Multiple-choice questions are questions that are followed by several answer choices. The correct answer is right in front of you. You just have to figure out which answer choice is the correct one. At times, you may have trouble figuring out the correct answer. If so, first try to get rid of the answers that you know are wrong. Then pick the best answer from the choices that are left.

Short-answer questions are questions that ask you to write out your answer. The answer might be a couple of words, or it might be a number. Whichever it is, be sure to use your best handwriting when you write your answer.

Open-response questions are questions that ask you to write out your answer and show your work (neatly!). These questions often have more than one part for you to answer. Remember to answer all parts of the question.

Now, here are the answers to the questions at the beginning of the chapter.

How did raw eggs fly hundreds of feet into the air?
In a recent science contest, high school students attached eggs to homemade rockets. They then launched the rockets into the air.

Just how big is a giant beach towel, anyway?
A giant beach towel that appeared at a festival in Spain measured 9.4 meters wide by 14.46 meters long. That's bigger than most classrooms. Imagine how many friends you could share this towel with!

What made Adam Sandler decide to become a performer?
When Adam was 17 years old, his brother convinced him to perform at a comedy club. He's been performing comedy ever since.

CHAPTER 2

Whole Numbers

How many stairs would you have to climb to reach the top of one mountain in Switzerland?

What are the Seven Summits?

What big, big contribution did Gutzon Borglum make to the United States?

A number is just a number, right? Wrong! There's more to a number than what you see at first glance. In this chapter you'll learn how to properly write, read, and compare numbers.

Numbers As Words

You can use **words** to write numbers. For example, you can write the numbers 1, 2, and 3 by using the words *one, two,* and *three*. The same goes for all numbers no matter how big or small. Here are some examples.

52 = fifty-two

49,781 = forty-nine thousand, seven hundred eighty-one

Place Value

Place value is the value of a digit based on its place in a number. For example, in the number 382, the digit 8 is in the tens place, so it has a value of 80. Every digit within a number has its own place value. Digits represent different amounts depending on where they appear in a number. The chart below shows the place values of all of the digits in the number 2,734,691.

millions	hundred thousands	ten thousands	thousands	hundreds	tens	ones
2	7	3	4	6	9	1

The number reads "two million, seven hundred thirty-four thousand, six hundred ninety-one."

Do you see how the words for each number match with the place values in the chart?

Comparing and Ordering Numbers

To arrange numbers in the correct order, you have to compare them. This means that you need to compare the digits in each place value of both numbers. Look at the numbers below. Can you tell which number is greater than the other?

56,783

56,692

To compare numbers, start with the largest place value first. In this case, the largest place value is the ten thousands (on the left). Both digits in the ten thousands place are fives, so you can't tell which number is greater. Now, compare the digits in the thousands place. Both digits in the thousands place are 6, so they are equal. You still can't tell which number is greater. Now, compare the digits in the hundreds place. The top number has a 7 in the hundreds place. The bottom number has a 6 in the hundreds place. The number 7 is greater than 6, so 56,783 is greater than 56,692. That's all there is to it!

Stairway to the Sky

Joker

Have you ever heard of a mountain that had stairs? It may sound weird, but a mountain in Switzerland has a flight of stairs running up its side. The stairway runs alongside a special train that goes up and down the steep mountain. If the train ever broke down, repair workers would climb the stairs to get to the train. They'd better hope the train never breaks down at the top of the mountain, though. To reach the top of the mountain, repair workers would have to climb 11,674 steps!

Directions: Answer the question based on the passage Stairway to the Sky.

▶ According to the passage, the mountain in Switzerland has 11,674 steps. Use words to write out the number 11,674.

Answer: _____

_____

Know It All Approach

First, read the question carefully and notice the important information. The question asks you to write out the number 11,674 in word form.

Now, answer the question. Read the number aloud. Pay close attention to the words you used to describe the number. Slowly think of the number again in your head as you write the words down.

Finally, double-check your answer. Go back and look at the number. Give place value names to each digit in the number. Make sure your place value names match up with the words you used when you wrote out the number in word form. You should write the words "eleven thousand, six hundred seventy-four."

Directions: Read the passages below and answer the questions that follow.

The Seven Summits

You may have heard of people climbing *one* of the world's tallest mountains, but how about all *seven* of them? Climbing all Seven Summits is the ultimate goal of many adventurous mountain climbers. The name *Seven Summits* refers to the tallest mountain peaks on each of the seven continents. As of March 2003, seventy-seven brave men and women may have made it to the top of all seven peaks. Atsushi Yamada climbed all Seven Summits when he reached the peak of Mount Everest, the tallest mountain in the world. He was only twenty-three years old when he completed his goal.

1. Hillary was making a time line of some of the important events in the history of Mount Everest exploration. She began her time line in the year 1841. She included the following dates on her time line:

   1856—Andrew Waugh measures Everest and finds it to be 29,002 feet high.
   1865—Peak XV is renamed Mount Everest.
   1852—Peak XV is said to be the world's highest mountain.
   1841—Sir George Everest first records Mount Everest's location.
   1854—Peak B is renamed Peak XV.
   1848—The British find that Everest's Peak B is 30,200 feet high.

   Of these dates, which would be the **last** date on the time line?

Answer: _____

2.

Official Height of Mount Everest As of November 1999
29,035

Which statement is true about the number 29,035?

○ A The digit in the ones place is even.
○ B There are 2 hundred thousands.
○ C The digit 9 is in the thousands place.
○ D The value of the 9 is 90,000.

Gutzon Borglum's Dream

Sculptors carve images for us to enjoy. The sculptor Gutzon Borglum dreamed of a sculpture many times larger than any statue. Borglum's dream was to honor the great presidents of the United States. He felt that such a large subject deserved a huge sculpture. That's why he sculpted Mount Rushmore!

The faces of George Washington, Thomas Jefferson, Theodore Roosevelt, and Abraham Lincoln are about sixty feet tall. That's about the height of a six-story building! It took fourteen years of hard work from 1927 to 1941 and cost nearly one million dollars to complete. Today, Mount Rushmore is one of America's most famous landmarks. In 2002, 2,159,189 visitors marveled at this national treasure.

3. How many people visited Mount Rushmore in 2002?

○ A two thousand, one hundred fifty-nine
○ B twenty-one thousand, five hundred ninety-one
○ C two million, fifteen thousand, nine hundred eighteen
○ D two million, one hundred fifty-nine thousand, one hundred eighty-nine

4. About how much did it cost to complete Mount Rushmore?

○ A $100,000
○ B $1,000,000
○ C $10,000,000
○ D $100,000,000

Subject Review

In this chapter you learned all about whole numbers. Review what you've learned so far.

You can write any number in **word form.** It's as easy as one . . . two . . . three million, seven hundred-thousand, six hundred fifty-nine!

Every digit within a number has a **place value** that tells how much it is worth. You can tell a digit's place value by where it stands within a number. Some examples of place values are ones, tens, hundreds, thousands, ten thousands, hundred thousands, and millions.

You can use place value to help you **compare and order numbers.** Compare the digit in the largest place value of all the numbers first. The largest place value is always on a number's far left side. If all of the numbers in that place value are the same, compare the digits in the next place value of all the numbers. You can do this when a question asks you to put numbers in order.

And now, here are the answers to the questions from page 21.

How many stairs would you have to climb to reach the top of one mountain in Switzerland?
A mountain near Spiez, Switzerland, has a stairway that has 11,674 steps.

What are the Seven Summits?
The Seven Summits are what mountain climbers call the seven highest mountain peaks, one on each of the seven continents. These mountains are Mount Everest (Asia), Mount Aconcagua (South America), Mount McKinley (North America), Mount Kilimanjaro (Africa), Mount Elbrus (Europe), Vinson Massif (Antarctica), and Carstensz Pyramid (Australasia). Many mountain climbers dream of climbing all seven peaks.

What big, big contribution did Gutzon Borglum make to the United States?
Gutzon Borglum was the sculptor who created one of America's most famous landmarks, Mount Rushmore. You can visit his sculpture in South Dakota.

CHAPTER 3
Computation and Word Problems

What makes a
hummingbird hum?

What is the oldest game
in the world?

Did Christopher Columbus
really discover the New
World?

Addition and Subtraction

Addition and **subtraction** are two skills that you'll use often in math. When you add and subtract, line up the numbers by place value. Then, add or subtract the numbers in each place value.

When the sum of the digits in one place value is greater than 9, you must **regroup.** Regrouping means naming a number differently. For example, the number 34 can be regrouped as 3 tens and 4 ones, or 2 tens and 14 ones. In the example below, 9 + 3 = 12, 12 is regrouped as 1 ten and 2 ones.

$$
\begin{array}{r}
\overset{1}{53} \\
+\,29 \\
\hline
82
\end{array}
$$

Regrouping is also necessary in subtraction problems. When the digit in the top number is less than the digit in the bottom number, you will need to regroup. In the example below, the number 772 is regrouped as 7 hundreds, 6 tens, and 12 ones.

$$
\begin{array}{r}
\overset{6\,12}{7\,\cancel{7}\,2} \\
-\,326 \\
\hline
446
\end{array}
$$

Check your answer in an addition problem by subtracting one of the numbers from the sum. You should get the other number. In the addition problem above, subtract 29 from 82 to get 53. Check your answer in a subtraction problem by adding the lesser number to the difference. You should get the greater number. In the subtraction problem above, add 446 and 326 to get 772.

Multiplication and Division

Multiplication and **division** are two other math skills that you'll need in order to solve many math problems. You have to know your multiplication facts to multiply and divide.

The example below shows how to multiply a two-digit number by a one-digit number. Place the greater number on top. Notice that $8 \times 2 = 16$, and 16 is a two-digit number. The 1 needs to be regrouped to the tens place. Then, the regrouped 1 is added to the product of 8×1.

Check your multiplication by dividing your answer by one of the factors in the multiplication problem. You should get the other factor. $96 \div 12 = 8$ or $96 \div 8 = 12$, so you know your answer is correct!

Now take a look at this division problem.

$$
\begin{array}{r}
21 \\
2\overline{)42} \\
\underline{-4\downarrow} \\
02
\end{array}
$$

2 divides into 4 twice and into 2 once. The answer is 21. Check your division by multiplying your answer by the divisor. The **divisor** is the number to the left (and outside) of the division bracket. When you multiply the answer by the divisor, you should get the number under the division bracket. In this case, $21 \times 2 = 42$, so you know your answer is correct!

Word Problems

Some math problems will be plain computation. Other math problems will be word problems. **Word problems** tell a story. You will have to read the word problems and figure out what operation to use to solve the problem.

The Bird That Hums

 If you've ever had a hummingbird buzz past your head, you might have mistaken it for a large bug at first. The hummingbird is the world's smallest bird. Some of these petite birds can weigh less than a penny! When it flies, the hummingbird makes a loud humming sound. The rapid flapping of the hummingbird's wings causes the sound. Small hummingbirds flap their wings 80 times per second. Larger hummingbirds flap their wings 10 times per second.

Directions: Answer the question based on the passage The Bird That Hums.

▶ A small hummingbird and a large hummingbird each flapped their wings for 5 seconds. How many times did the two hummingbirds flap their wings in 5 seconds altogether?

- ○ A 50
- ○ B 80
- ○ C 400
- ○ D 450

Know It All Approach

First, read the question carefully and find the important information. A small hummingbird can flap its wings 80 times per second. A large hummingbird can flap its wings 10 times per second. You need to figure out how many times the birds flapped their wings in 5 seconds.

Together, the birds flapped their wings 80 + 10 = 90 times in 1 second. Multiply by 5 to find out how many times they flapped their wings in 5 seconds. 90 × 5 = 450 times.

Now read the answer choices and bubble in the correct answer choice, (D). If you had trouble finding the answer, you could get rid of wrong answer choices. The small hummingbird would flap its wings 80 × 5 = 400 times in 5 seconds. The answer must be greater than 400 because you must add the number of times that a large hummingbird flaps its wings. Get rid of answer choices (A), (B), and (C). You are left with (D).

Directions: Read the passages below and answer the questions that follow.

How Old?

Mancala may be the oldest game in the world. Experts believe that people first began playing Mancala around 1400 B.C. Two players divide up 48 playing stones equally. The object of the game is to be the first player to get the playing stones into your *kalaha.* The kalaha is a large bin on the playing board. There are many strategies; it may be as complicated as chess.

1. Franny needs to set up her side of the board to play Mancala. There are 6 bins on her side of the board. If she divides her 24 playing stones equally between her 6 bins, how many playing stones should Franny place into each bin?

Show your work.

Answer: _____ playing stones

Rewrite That History Book

Did Christopher Columbus discover America? No way. Native Americans already lived throughout the "New World" when Columbus first landed in the Bahamas. He did, however, lead the first expedition to the Americas from the European mainland.

2. Columbus came over on three ships: the Niña, Pinta, and Santa Maria. The Niña had 24 crewmembers. The Pinta had 26 crewmembers. The Santa Maria had 40 crewmembers.

Part A

If all the crewmembers were divided evenly amongst the three ships, how many crewmembers would be on each ship?

Answer: _____

Part B

Describe how you would explain to a friend how to solve this problem. You may use words, numbers, or pictures in your answer.

Answer: _____

Subject Review

You learned about **addition, subtraction, multiplication,** and **division** in this chapter. You also learned how to solve **word problems.**

When you add and subtract, you have to remember to **regroup** your numbers when necessary. To check your answer to an addition problem, just subtract from the bottom up. To check your answer to a subtraction problem, just add from the bottom up.

When you multiply, you have to remember to regroup your numbers when necessary. To check your answer to a multiplication problem, just divide your answer by one of the numbers in the problem.

When you divide, you have to remember to drop digits down from the number under the division bracket when necessary. A **divisor** is the number to the left (and outside) of the division bracket. To check your answer to a division problem, just multiply the divisor by your answer.

Below are answers to the questions from the start of the chapter.

What makes a hummingbird hum?
The fast beating of a hummingbird's wings produces the humming sound. Some hummingbirds beat their wings 80 times per second!

What is the oldest game in the world?
Some experts think that a West African and Nigerian game called Mancala is the oldest game in the world. People began playing the game around 1400 B.C.

Did Christopher Columbus really discover the New World?
No way! Native Americans already lived throughout North America by the time Columbus landed in the Bahamas.

CHAPTER 4
Operations and Properties of Numbers

How many matchsticks does it take to build a violin?

How can you stop your parrot from making splotches on the furniture?

What is a sari and how long is a really long one?

You're doing great so far! In this chapter you'll learn about operations and different properties of numbers. This is a fancy way to say that you'll learn more about how numbers work together.

Operations

An **operation** is a rule you follow to compute numbers. The great news is that you already know the four basic operations. Addition, subtraction, multiplication, and division are all operations. You can use these operations to find answers to math problems, especially word problems. Take a quick look at each operation.

In **addition,** you combine numbers, such as $4 + 5 = 9$. Words that may tell you to add are *total, sum, together,* and *in all.*

In **subtraction,** you take one number away from another number. You can use subtraction to check an addition problem. To make sure that $4 + 5 = 9$, for example, you could use one of the subtraction problems below. Don't forget to use addition to check your subtraction.

$$9 - 5 = 4$$
$$9 - 4 = 5$$

Words that may tell you to subtract are *difference* and *less than.*

Multiplication is a way to write repeated addition. In the problem $3 \times 3 = 9$, you are finding three groups of 3. This is the same thing as adding three 3s ($3 + 3 + 3 = 9$). Multiplication is a faster way to do repeated addition. Use multiplication to solve word problems if you are finding many groups of the same number.

Division is a way to find the number of groups in another number. The division problem $15 \div 3 = 5$ shows that there are 3 groups of 5 in the number 15. To check this answer, multiply the answer, 5, by the divisor, 3. You should end up with the first number. $5 \times 3 = 15$. Think of division as the opposite of multiplication. In multiplication, you are making many groups. In division, you are breaking up a number into many equal groups.

Properties of Operations

Mathematical operations have different properties. Three important properties are the commutative property, the associative property, and the distributive property. These are big words, but you won't be so scared of them after you read the descriptions below!

The **commutative property** means that you can change the order of the numbers and it won't change the answer. This property works only with addition or multiplication.

$3 + 7 = 10$ is the same as $7 + 3 = 10$

$2 \times 8 = 16$ is the same as $8 \times 2 = 16$

The **associative property** means that the answer will not change no matter how you group numbers in a problem. Parentheses show how numbers are grouped. This property works only with addition or multiplication.

$(3 + 6) + 8 = 17$ is the same as $3 + (6 + 8) = 17$

$(2 \times 4) \times 3 = 24$ is the same as $2 \times (4 \times 3) = 24$

The **distributive property** applies to problems with multiplication *and* addition. When you multiply a sum in parentheses, you can multiply each number in the sum separately, and then add the products.

$5(3 + 4) = 5(7) = 35$ is the same as $(5 \times 3) + (5 \times 4) = 15 + 20 = 35$

$8(1 + 6) = 8(7) = 56$ is the same as $(8 \times 1) + (8 \times 6) = 8 + 48 = 56$

A Matchstick Violin

Jack Hall was a sailor in the 1930s. He was onboard a ship called the *Eastwick* when he thought of an unusual way to pass the time. He began collecting used wooden matchsticks that his fellow sailors tossed away. At first Jack wasn't sure what he would do with the matchsticks. Then he decided to build musical instruments with them. His first creation was a full-size, working violin and bow. It took 14,000 matchsticks to build his first instrument!

Directions: Answer the question based on the passage A Matchstick Violin.

▶ Neema needs 98 wooden matchsticks to complete her model of a bongo drum. She has 23 matchsticks so far. Which operation should Neema use to figure out how many more matchsticks she needs?

$$98 \boxed{?} 23 =$$

○ A addition
○ B subtraction
○ C multiplication
○ D division

Know It All Approach

Read the question carefully. The important numbers are 98 and 23. The important words are *which operation* and *how many more*. Think of an operation that helps you figure out how many more matchsticks Neema needs.

Neema needs 98 total matchsticks and she has 23. To figure out how many *more* matchsticks she needs, you must subtract 23 from 98. The question asks you to choose which operation belongs in the box, so the answer must be subtraction.

Retrace your steps and answer the question a second time. In this case, reread the question carefully to make sure you understood it. Finally, fill in the bubble next to the correct answer choice. In this case, bubble in (B).

Directions: Read the passages below and answer the questions that follow.

No More Bird Droppings!

In the late 1990s, several people invented a bird diaper. The bird diaper allows a pet bird to fly around its owner's house without getting droppings on the furniture. Imagine what some houses must have looked like before the bird diaper came along. Gross!

1. Vito is starting his own bird diaper business. He has made 504 diapers so far and wants to distribute them evenly among his 12 customers. He wrote the following:

    $$504 \div 12 = 42$$

    Which expression should Vito use to see if his number sentence is correct?

    ○ A 42×12
    ○ B $42 + 12$
    ○ C $42 \div 12$
    ○ D $42 - 12$

2. Vito plans to work seven days a week and make four diapers a day. He uses the expression 7×4 to figure out how many bird diapers he'll make every week.

    Which expression is *not* the same as 7×4?

    ○ A $4 + 4 + 4 + 4 + 4 + 4 + 4$
    ○ B $7 + 7 + 7 + 7$
    ○ C 4×7
    ○ D $7 \times 7 \times 7 \times 7$

One Long Sari!

Asian women often wear flowing robes called saris. In 1998, one person made a 396-foot long sari! It was decorated with paintings of famous women.

3. To describe how long the sari was, Quentin wrote this number sentence:

    $$2 \times 3 \times 66 = 66 \times 3 \times 2$$

    Which property tells you that the number sentence is correct?

Answer: _____

Subject Review

An **operation** is a rule you follow to change pairs of numbers. The four types of operations are addition, subtraction, multiplication, and division. You can use these operations to find answers to math problems.

Operations are related, so you can check your answers by using other operations.

- To double-check an addition problem, just subtract one of the numbers to be added from the sum.
- To double-check a subtraction problem, add the difference to the smaller number that was subtracted.
- To double-check a multiplication problem, divide the product by one of the factors.
- To double-check a division problem, multiply the quotient (answer) by the divisor.

Mathematical operations have different properties.

- The **commutative property:** When you add or multiply, changing the order of the numbers won't affect the answer.
- The **associative property:** No matter how you group numbers in an addition or multiplication problem, the answer will always be the same.
- The **distributive property:** When you are multiplying a sum in parentheses, you can multiply each number in the sum separately. Then, you add the products.

Now, here are the answers to the earlier questions.

How many matchsticks does it take to build a violin?
In the 1930s, Jack Hall made a working violin and bow with 14,000 matchsticks.

What can you do to stop your parrot from making splotches on your furniture?
A group of people invented a bird diaper that would prevent birds from making a mess as they flew freely around their owners' houses.

What is a sari and how long is a really long one?
A sari is a traditional Asian robe. A 396-foot-long sari was created in 1998.

Know It All! Elementary School Math

CHAPTER 5
Decimals

How much did a loaf of
bread cost in 1962?

Who banned the Olympic
Games?

What's so special about
the Rochester Red Wings
and the Pawtucket Red
Sox?

Decimals

A **decimal** is a number that shows part of a whole. Decimals are written with a symbol called a **decimal point.** The decimal point separates the whole numbers from the digits that stand for parts of a whole number. When there are ten equal parts, each part is called a tenth. When there are one hundred equal parts, each part is called a hundredth.

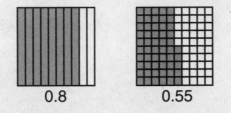

0.8   0.55

Each figure above represents one whole. The figure on the left shows a whole divided into tenths with eight parts shaded. This represents eight-tenths. The figure on the right shows a whole divided into hundredths with fifty-five parts shaded. This represents fifty-five hundredths.

Adding and Subtracting Decimals

Adding and subtracting decimals is like adding and subtracting whole numbers. Line up the decimal points in the numbers and add or subtract as usual. Then, remember to include the decimal point in your answer.

You use decimal points when you add and subtract sums of money. Pretend that you broke open your piggy bank and found $5.25 inside. Your neighbor just paid you $1.25 to pull weeds in his garden. (Your neighbor is *mega*-cheap.) To figure out how much money you have in total, you must add. Notice how the decimal points are lined up in the problem. See also how the answer includes a decimal point.

$$
\begin{array}{r}
\overset{1}{\$5.25} \\
+\ \$1.25 \\
\hline
\$6.50
\end{array}
$$

Bread Rising

Everyday items cost much more today than they did years ago. This increase in the cost of everyday items is called inflation. Back in 1962, a loaf of bread cost only $0.32. In 2002, a loaf of bread cost $1.26. That's an increase of $0.94 in forty years. Imagine how much a loaf of bread will cost two hundred years from now!

Directions: Answer the question based on the passage Bread Rising.

▶ At a garage sale, Theo found an old grocery store sign that read

Part A

What is the **least** expensive way to buy 4 loaves of bread according to the sign?

   4 separate loaves of bread   or   a 4-pack of bread loaves

Show your work.

Answer: _____

Part B

What is the **least** amount of money you would need to buy exactly 10 loaves of bread according to the sign?

Show your work.

Answer: $_____

Know It All Approach

This open-response question has more than one part. Remember to answer both part A and part B. First, read the question carefully and notice the important words and information. Be sure to study the picture too. In the picture and part A, the important words and numbers are *$0.32 per loaf, 4 loaves for $1.00,* and *least expensive way.* In part B, the important words and numbers are *least amount* and *10 loaves.*

Next, figure out the answers. Part A asks you to decide which is the least expensive way to buy bread—in 4 separate loaves or by the 4-pack. Multiply $0.32 × 4. It would cost you $1.28 to buy 4 separate loaves. The sign says that a 4-pack of loaves costs $1.00. $1.00 is less than $1.28, so buying the 4-pack is the cheapest way to go.

Part B asks you to find the least amount of money you'd need in order to buy 10 loaves of bread. Two 4-packs make 8 loaves, and would cost $2.00 total. You need 2 more loaves to make 10. Two more loaves is $0.32 cents × 2 = $0.64. $2.00 + $0.64 = $2.64. So, $2.64 is the least amount of money you'd need in order to buy 10 loaves of bread.

Review your steps for both parts of the question. Reread the question and make sure that your numbers match the word problem. Then, review your work to make sure you did the math correctly. Finally, use your best handwriting to write the answers to the question.

Directions: Read the passages below and answer the questions that follow.

Olympic Games

Today's Olympics are very different from the ancient Olympic Games. The 2000 games in Sydney, Australia, drew more than 10,000 athletes from almost 200 countries. These athletes competed in 300 events.

The ancient Olympic Games began more than 4,000 years ago in Greece. At first, these Games had only one race. The competitors were all male and from Greece. In fact, the spectators were all male, too, because women were not allowed into the stadium.

The Roman Empire conquered ancient Greece, but that didn't stop the Games. They just moved to Rome, Italy. The Roman Emperor Theodosius I banned the Games in A.D. 394. The Games may have still continued at the original site, but Theodosius later sent a Roman army to destroy the temples and arena.

1. In the 2000 Olympic Games, U.S. Olympic Team member Stacy Dragila took part in the pole vault competition. She jumped over a bar that was 4.6 meters high.

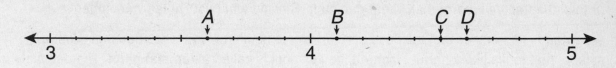

Which arrow on the number line points to the height that Stacy Dragila jumped?

○ A A
○ B B
○ C C
○ D D

A Minor League (Almost) Quadruple-Header

In 1981, two minor league teams—the Rochester Red Wings and the Pawtucket Red Sox—made baseball history. The two teams played a game that lasted a total of 33 innings over the course of two days. One game typically has only 9 innings, so this game was almost four games long!

2. Wendy attended a recent Pawtucket Red Sox game. She had $10 to spend at the stadium gift shop. She bought two of the items below and received $3.50 in change.

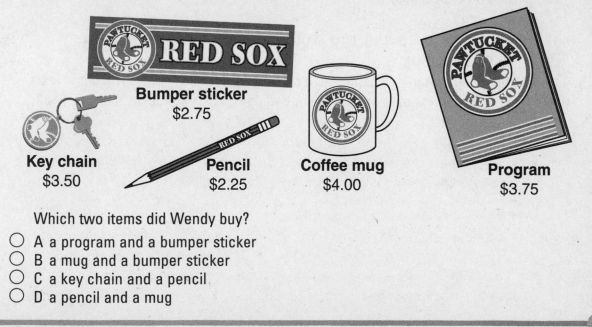

Bumper sticker
$2.75

Key chain
$3.50

Pencil
$2.25

Coffee mug
$4.00

Program
$3.75

Which two items did Wendy buy?

○ A a program and a bumper sticker
○ B a mug and a bumper sticker
○ C a key chain and a pencil
○ D a pencil and a mug

Subject Review

In this chapter, you learned all about decimals. Review what you've learned so far.

A **decimal** is a number that includes parts of whole numbers. You write decimals with a symbol that looks like a period. In math, that symbol is called a **decimal point**. The decimal point has numbers after it that stand for parts of a whole. The numbers after the decimal point have place values (tenths, hundredths, and so on), just as whole numbers do.

When doing a math problem that involves decimals, always make sure to line up the decimal points in the problem. Also, remember to include the decimal point in your answer.

Decimal points are used in math problems that involve money. They are also used to describe measurements with metric units.

Remember the questions from the beginning of the chapter? Here are the answers.

How much did a loaf of bread cost in 1962?
A loaf of bread cost $0.32 in 1962.

Who banned the Olympic Games?
A Roman emperor named Theodosius banned the Olympic Games in ancient Greece. The modern Olympic Games began in 1896, almost 1,500 years after the end of the ancient Games.

What's so special about the Rochester Red Wings and the Pawtucket Red Sox?
These two minor league teams played a really long baseball game. The game lasted a total of 33 innings over the course of two days. The Pawtucket Red Sox finally won the game 3 to 2.

Directions: Read the passages below and answer the questions that follow.

Keep Your Eyes to Yourself!

 A woman from Chicago named Kim Goodman has an unusual talent, to say the least. She can pop her eyeballs out 0.43 inches past her eye sockets. And you thought this kind of thing happened only in cartoons!

1. Nathan, Vera, Chanda, and Kenji are having their own eyeball-popping contest. The table below shows how far they can pop their eyeballs past their eye sockets.

Eyeball Popping Distances

Contestant	Distance Popped in inches
Chanda	0.33
Kenji	0.40
Nathan	0.42
Vera	0.35

Part A

Which contestant came the closest to popping his or her eyeballs $\frac{1}{2}$ inch?

Answer: _____

Part B

Which contestant popped his or her eyeballs farther than Chanda but not as far as Kenji?

Answer: _____

2. Nathan, Vera, Chanda, and Kenji decided to have three eyeball-popping contests. They are selling tickets to all three eyeball-popping contest showings. Their ultimate goal is to sell a total of 150 tickets. They have already had two shows. They sold 34 tickets to the first show and 73 tickets to the second show. How can they find out how many more tickets they need to sell?

○ A 150 − (73 − 34)
○ B 150 + (34 − 73)
○ C 150 − (34 + 73)
○ D 150 + (34 + 73)

3. Imagine you are buying a ticket to the third show of the eyeball-popping contest. Tickets are selling for $3.75. If you hand the cashier $5.00, which picture shows the amount of change you should get back?

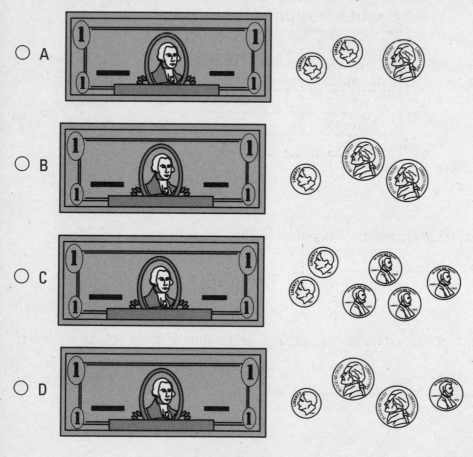

Surin Soccer

Have you ever heard of elephants playing soccer? Well, that's exactly what happens at the Surin Elephant Roundup in Thailand. The Surin Elephant Roundup is a festival held every year to honor elephants. The festival was first held in 1960. It has since become a very popular tourist attraction.

One of the main events at the festival is the elephant soccer game. People ride on the backs of the elephants to direct them in the game. The rest is up to the elephants. The huge creatures pick up the soccer ball with their trunks and drop-kick it across the field. Every so often, one of them scores a goal. Can you imagine if you were a goalie who had to block a ball that an elephant kicked? Forget it!

4. Tai is setting up rows of chairs at the Surin Elephant Roundup. Some rows have an odd number of chairs, and some rows have an even number of chairs. When added together, which kind of numbers would give Tai a total that is even?

- ○ A odd + even
- ○ B odd + odd + odd
- ○ C even + even + odd
- ○ D odd + odd

5. Tai counted 432 people watching the elephants play soccer. He wrote the following number sentence to describe the number of people in the audience:

$$4 \times 9 \times 12 = 9 \times 12 \times 4$$

Is this number sentence correct? Explain how you know whether this number sentence is correct without having to do the math.

_____

_____

_____

_____

6. For a few days, Tai worked at one of the festival's beverage stands. The numbers below show how many baht Tai's stand made each day he worked. (Baht is the name for Thai money.) Write the numbers in order from least to greatest.

25,463   25,649   26,942   24,950

7. Tai helped to clean the elephants' stalls one day. His boss asked him to collect the tools in the storage room. Tai found 5 brooms. He found 8 more shovels than brooms. He found 4 fewer rakes than shovels. How many rakes did Tai find?

○ A 5
○ B 7
○ C 9
○ D 13

CHAPTER 6

Fractions

How does a spider land a good acting role in a movie?

How long has pizza been around?

Just how big can a sand sculpture get?

Writing Fractions

A **fraction** is a number that shows parts of a whole. The **denominator** is the bottom number of a fraction. This shows the number of equal parts that a whole is broken up into. The figure below shows a pizza divided into 4 equal slices. The denominator is 4.

The **numerator** is the top number of a fraction. This shows the number of parts that you are interested in. The figure below shows that only one slice of the pizza is left. The fraction that shows how much pizza is left is $\frac{1}{4}$.

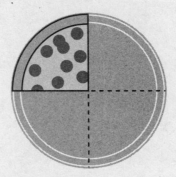

Comparing Fractions

Sometimes you will need to figure out when one fraction is greater than or less than another fraction. When you compare two fractions that have the same denominator (bottom number), you just have to compare the numerators (top numbers). The whole figures below are broken up into fifths. One figure has two parts shaded and represents $\frac{2}{5}$. The other figure has one part shaded and represents $\frac{1}{5}$. Two-fifths is greater than one-fifth, or $\frac{2}{5} > \frac{1}{5}$.

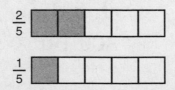

Know It All! Elementary School Math

When fractions have different denominators, you can't just compare the two numerators. The pieces are different in size, so it's not always clear which fraction is greater. You can use pictures to help compare fractions with different denominators. For example, look at the figures below to compare the fractions $\frac{1}{2}$ and $\frac{3}{4}$. Just remember to keep the whole shapes equal in size!

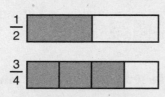

From the picture, you can tell that $\frac{3}{4}$ is greater than $\frac{1}{2}$.

Adding and Subtracting Fractions

When you add and subtract fractions with the same denominator, you are really adding or subtracting the numerator. See the figure below.

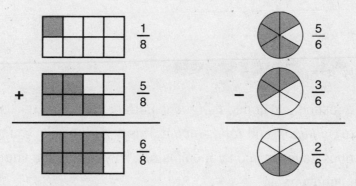

Notice that you add or subtract the numerators only, but leave the denominators alone. This makes sense because the number of parts that the whole is broken up into does not change.

Sometimes you may have to add or subtract fractions and whole numbers. Just calculate the fraction part, and then calculate the whole-number part.

$$1\frac{3}{8} + 2\frac{1}{8} = 3\frac{4}{8}$$

Give That Spider a Job!

Have you ever heard of a talent manager for spiders? Well, that's exactly what Steven Kutcher does for a living. Kutcher is a bug expert who finds "acting" parts for bugs in movies and television shows. One of Kutcher's spiders played a key role in the popular movie *Spider-Man*. The spider even wore makeup! Kutcher had to paint the spider red and blue for its role in the film.

Directions: Answer the question based on the passage Give That Spider a Job!

▶ If Steven Kutcher painted $\frac{1}{8}$ of a spider blue and $\frac{4}{8}$ of the same spider red, what fraction shows the total amount of the spider that was painted?

- ○ A $\frac{5}{16}$
- ○ B $\frac{3}{8}$
- ○ C $\frac{1}{8}$
- ○ D $\frac{5}{8}$

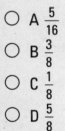

Know It All Approach

Carefully read the question and notice that the important numbers are $\frac{1}{8}$ and $\frac{4}{8}$. The important words are *additional* and *total amount*. This question asks you to add the two fractions. The fractions have common denominators, so you can just add the numerators and keep the same denominator. $\frac{1}{8} + \frac{4}{8} = \frac{5}{8}$.

Check your work by retracing your steps and answering the question a second time. Make sure you added correctly. Also, make sure you remembered to add just the numerators together and **not** the denominators. Now, read the answer choices. $\frac{5}{8}$ matches answer choice (D)! Carefully fill in the bubble for answer choice (D).

If you had trouble coming up with an answer, try to get rid of answer choices that you know are wrong. For example, answer choice (A) is wrong because the denominators were added. (Don't do that!) Answer choice (B) is wrong because it shows the subtraction of $\frac{1}{8}$ from $\frac{4}{8}$, rather than addition. And answer choice (C) is too small!

Directions: Read the passages below and answer the questions that follow.

Ancient Food

No one knows exactly when pizza was invented. However, experts believe that the word *pizza* dates back to around A.D. 1000 in Rome or Naples, Italy. The word meant pie and just described meat on a flat piece of bread.

1. Look at the figure below of two pizzas. According to the figure, which of the following statements is true?

   ○ A $\frac{4}{8} > \frac{2}{4}$

   ○ B $\frac{4}{8} = \frac{2}{4}$

   ○ C $\frac{2}{4} > \frac{4}{8}$

   ○ D $\frac{4}{8} < \frac{2}{4}$

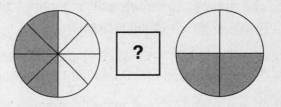

Sand Mania

In 1991, thousands of people created a huge sand sculpture in Myrtle Beach, South Carolina. When completed, the detailed sand sculpture stretched more than 86,000 feet across the beach!

2. Kyle is building a sand castle. He has just filled a bucket with sand. The bucket holds $3\frac{2}{3}$ cups of sand. He then removed $1\frac{1}{3}$ cups of sand from the bucket to make a tower for his sand castle. How much sand remains in the bucket?

   ○ A $4\frac{1}{3}$

   ○ B $2\frac{1}{6}$

   ○ C $2\frac{1}{3}$

   ○ D $\frac{1}{3}$

Subject Review

This chapter was all about fractions.

A **fraction** is a number that stands for part of a whole. The **denominator** is the bottom number of a fraction. It tells you the number of equal parts. The **numerator** is the top number of a fraction. It tells you the number of parts that you are interested in.

When you compare two fractions that have the same denominator, just compare the two numerators. The fraction with the greater numerator is greater than the other fraction.

When fractions have different denominators, you can't just compare the two numerators. You can draw pictures representing each fraction and compare the pictures to help you determine which fraction is greater.

When you add fractions with common denominators, just add the numerators and leave the denominators alone. Do **not** add the denominators.

When you subtract fractions with common denominators, just subtract the numerators and leave the denominators alone. Do **not** subtract the denominators.

Here are the answers to the earlier questions.

How does a spider land a good acting role in a movie?
Steven Kutcher, a bug expert, trains spiders, butterflies, and insects for parts in movies, commercials, and television shows.

How long has pizza been around?
No one knows for sure, but the word *pizza* has been around since at least the year 1000. The experts also debate where pizza originally came from.

Just how big can a sand sculpture get?
A sand sculpture created in Myrtle Beach, South Carolina, was more than 86,000 feet long.

CHAPTER 7

Fractions, Decimals, and Percents

What is the biggest and fastest mammal in the sea?

Tiddlywinks? What's that?

What did a group of teenagers discover in Spring Valley, Minnesota?

There's more to learn about fractions and decimals. You need to learn the relationship between fractions and decimals—and percents, another way to show parts of a whole.

Percents

A **percent** shows a part of 100. Percent is represented by the symbol %. Imagine a parking lot filled with 100 bicycles. If 20 of those bicycles are silver, then 20% of the bicycles are silver. Imagine a book with 100 pages. If 99 of those pages have pictures, then 99% of the pages have pictures.

Converting Fractions, Decimals, and Percents

Fractions, decimals, and percents are all ways of showing parts of a whole. Think of them as different ways of writing the same thing. Use the chart below to see how fractions, decimals, and percents are related.

Fraction	Decimal	Percent
$\frac{1}{10} = \frac{10}{100}$	0.10	10%
$\frac{1}{5} = \frac{20}{100}$	0.20	20%
$\frac{1}{4} = \frac{25}{100}$	0.25	25%
$\frac{1}{2} = \frac{50}{100}$	0.50	50%
$\frac{3}{4} = \frac{75}{100}$	0.75	75%
$\frac{4}{4} = \frac{100}{100}$	1.00	100%

Percents are parts out of 100. You can rewrite them as fractions out of 100. Take the number part of the percent and write that over 100 to get the fraction. Some examples are $70\% = \frac{70}{100}$ and $16\% = \frac{16}{100}$.

When a fraction has a denominator of 100, you can convert it to a percent. Just write the numerator with a percent sign. If the fraction has a denominator of 10, multiply the numerator by 10 before writing the percent sign. For example, $\frac{43}{100} = 43\%$ and $\frac{3}{10} = 30\%$.

Percents and decimals are converted by moving the decimal points two places to the left or right. For example, 89% = 0.89, 60% = 0.60, and 0.45 = 45%. This will be explained in more detail on the next page.

Finally, fractions out of 100 or 10 can be converted to decimals by writing the numerator with a decimal point. Some examples are $\frac{30}{100} = 0.30$, $\frac{55}{100} = 0.55$, and $\frac{4}{10} = 0.40$.

Know It All! Elementary School Math

Calculating with Percents

To calculate with percents, you need to rewrite the percent as a decimal before performing the operation. To find 40% of 80, first rewrite the percent as a decimal: 40% = 0.40. Then multiply the decimal by 80. 0.40 × 80 = 32, so 40% of 80 is 32.

Step 1	Step 2
40% = 0.40 Change the percent to a decimal.	80 × 0.40 ──── 00 + 3200 ──── 32.00 Multiply the whole number by the decimal number.

Every number has a decimal point, even if you don't see it. For example, 40% is the same as 40.0%. And 32 is the same as the 32.00 shown above. So if you have to change a percent to a decimal, just imagine a decimal point to the right of the last percent number. Then move it two spaces to the left. This way, 40% becomes 0.40 and 5% becomes 0.05.

Once in a Whale. . .

The blue whale is the largest marine mammal. It grows to more than 100 feet long and weighs around 200 tons. If you guessed that such a large creature would move very slowly, you'd be wrong. In fact, the blue whale is the fastest mammal in the sea. It can swim up to 30 miles per hour!

Directions: Answer the following question based on the passage Once in a Whale.

▶ After studying sea life with his class, Mr. Schwartz asked his students to name their favorite sea creatures. Here are the results of his survey.

Favorite Animal	Number of Students
Shark	25
Blue Whale	10
Octopus	5
Total	**40**

What percent of the students liked the blue whale the most?

Answer: _____

Know It All Approach

This is a short-response question. The question has a chart, so study it carefully. Then look for words and information that you will need in order to figure out the answer. The important words in the question are *What percent of the students* and *blue whale*. The important words and numbers in the chart are *Blue whale, 10, Total,* and *40.*

The question asks you to figure out what percent of students liked the blue whale best. According to the chart, 10 out of 40 total students liked the blue whale best. With these numbers, you can make the fraction $\frac{10}{40}$. The simplest version of this fraction is $\frac{1}{4}$. Now look at the fraction conversion chart on page 56. According to the chart, the fraction $\frac{1}{4}$ is the same as 25%.

Make sure you read the question and the chart correctly. Also make sure you did your work correctly before writing the answer on the line provided.

Directions: Read the passages below and answer the questions that follow.

Wink Wink

The game of tiddlywinks dates back to the mid- to late 1800s. The game requires players to flip playing chips called winks into a cup. The players flip the winks into the air by pressing down firmly on the edge of a wink with another chip.

Today, the game of tiddlywinks must compete with modern forms of amusement like video games. Many tiddlywinks fans still exist, however. In 1995, a man in England set a record for the farthest wink flipped. The wink flew 31 feet and 3 inches!

Know It All! Elementary School Math

1. Jabar, Denise, Michael, and Ike are all learning to play tiddlywinks. They had a contest to see how far they could flip their winks. The table below shows how far each of the winks flew.

Wink Flipping Distances

Player	Distance Flipped in Inches
Jabar	1.25
Denise	0.75
Michael	0.45
Ike	0.65

Which player flipped his or her wink the closest to $\frac{1}{2}$ inch?

Answer: _____

2. Ike had 10 winks total. He flipped 6 of his winks into the cup. What percent of his winks did he flip into the cup?

- ○ A 6%
- ○ B 60%
- ○ C 40%
- ○ D 0.06%

A Remarkable Discovery

In 1995, a group of teenage cave explorers and their adult guide made a remarkable discovery. While exploring a cave in Spring Valley, Minnesota, they discovered an unexplored cavern. To reach the cavern, the group had to crawl through a narrow tunnel that was 50 feet long. The tunnel was so narrow that one of the young cavers got stuck at one point. His friends had to help him get free.

3. Amy is saving her money so that she can go on a cave exploration trip. The trip costs $100. She has saved $40 so far. What percent of $100 has Amy saved?

- ○ A 400%
- ○ B 0.4%
- ○ C 4%
- ○ D 40%

Subject Review

In this chapter, you learned about percents and how to convert between fractions, decimals, and percents. A **percent** is the same as a fraction with a denominator of 100. For example, $\frac{20}{100} = 20\%$.

Fractions, decimals, and percents are all numbers that show parts of a whole. They are basically different ways of writing the same thing. For example, $\frac{50}{100} = 0.50 = 50\%$.

All numbers can be written with a decimal point, but sometimes the decimal point is not shown if there are only zeros after it. For example, 60.0% is the same as 60%.

When you're asked to change a percent to a decimal, just imagine a decimal point to the far right of the percent. Then move it two spaces to the left. 20% and 0.20 stand for the same amount.

Now, here are answers to the questions from the beginning of the chapter.

What is the biggest and fastest mammal in the sea?
The blue whale is the biggest and fastest marine mammal.

Tiddlywinks? What's that?
Tiddlywinks is a game that people started playing in the mid- to late 1800s. The object is to flip chips called winks into a cup.

What did a group of teenagers discover in Spring Valley, Minnesota?
A group of teenage cave explorers discovered an unexplored cavern.

CHAPTER 8

Patterns

What does Maurice Bennett do with his burnt toast?

What kind of weird machine did Martin Goetze invent?

How many instruments is it humanly possible to play at once?

Patterns

A **pattern** is a series of numbers or shapes that follows a rule.

Some math problems ask you to make predictions based on a pattern. Look at the pattern of numbers below.

2   7   12   17   22

Do you see the pattern? What could you add to each number to get the next number? When you add 5 to each number in the line, the sum equals the number to its right. When you add 5 to 2, you get 7. When you add 5 to 7, you get 12, and so on. Can you predict what number will come after 22? If you answered 27, you're correct. $22 + 5 = 27$.

There are patterns in many real-life situations. See if you can figure out the pattern in the paragraph below.

▶ Renah works at the supermarket. She is creating a display to promote the sale of vegetable soup. She decides to stack the cans of soup into four rows. In the first row, Renah uses 10 cans. In the next row, she uses 20 cans. In the third row, she uses 40 cans. If Renah continues the pattern, how many cans of soup will she use in the fourth row?

To answer this question, identify the pattern in the numbers.

10   20   40

Can you see the pattern? In each row, the number of cans doubles. $10 \times 2 = 20$ and $20 \times 2 = 40$. So, to figure out how many cans Renah uses in the fourth row, simply double 40. $40 \times 2 = 80$. Renah will use 80 cans in the fourth row.

Sometimes a series of shapes will form a pattern. Look at the shapes below. What do you think the next shape in the line will look like?

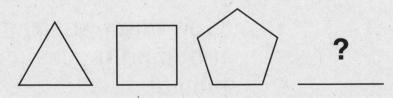

Compare the shapes of the figures. What is different about each shape? If you said that each shape has a different number of sides, you'd be right. The first shape has 3 sides. The second shape has 4 sides. The third shape has 5 sides. The pattern 3, 4, 5 suggests that the next shape will have 6 sides. It will probably look like this.

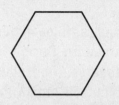

The Toastman

 Maurice Bennett, an artist from New Zealand, uses unusual materials for his art. He makes mosaics (tile pictures) out of—get this—burnt toast. Bennett uses different shades of burnt toast as a painter would use different colors of paint. He has even created a burnt-toast portrait of Elvis Presley!

Directions: Answer the question based on the passage The Toastman.

▶ Carl decided to make his own art out of burnt toast. He used dark and light pieces of toast to make the pattern below.

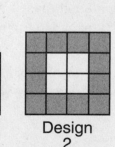

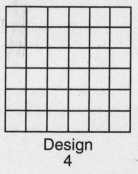

Design 1   Design 2   Design 3   Design 4

Carl would like to continue the same pattern to make Design 4. How many pieces of dark toast should he use in his fourth design?

○ A 16
○ B 18
○ C 20
○ D 32

Know It All Approach

This multiple-choice question asks you to predict what the pattern will do next, so study the pictures carefully.

What do you notice about the dark squares in each design? Only the squares around the border are dark. Count them to find a pattern. Design 1 has 8 dark squares. Design 2 has 12 dark squares. Design 3 has 16 dark squares. Think about the numbers 8, 12, and 16. Do you notice a pattern? What do you get if you add 4 to 8? 8 + 4 = 12. Now add 4 to 12. The answer is 16. It looks like you can add 4 to each design to get the number of dark squares. To figure out how many dark squares Design 4 will have, just add 4 to 16. The answer is 20.

Double-check your answer and read all of the answer choices. Answer choice (C), 20, is correct. Make sure none of the other answer choices could be correct, and fill in the correct bubble.

Directions: Read the passages below and answer the questions that follow.

Dimpler

An inventor from Germany, Martin Goetze, invented a dimple-making machine in 1896. The machine was a big clamp with a small, vibrating ball on it. If you wanted a dimple on your cheek, you could clamp the machine onto your head and place the little ball wherever you wanted your new dimple to be. There's probably a good reason that you can't find these dimple-making machines at the local department store!

1. Victor imagines that he could make a lot of people happy if he started a dimple-making business. Victor predicts that the number of customers would increase according to a pattern shown below.

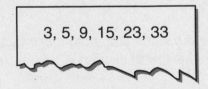

3, 5, 9, 15, 23, 33

What number comes next in the sequence?

Answer: _____

2. Victor plans to charge more money per dimple as his dimple-making business becomes more popular. On the first day, he will charge 1 quarter per dimple. On the second day, he will charge 3 quarters per dimple. On the third day, he will charge 9 quarters per dimple. How many quarters will Victor charge on the fifth day?

- ○ A 13
- ○ B 18
- ○ C 27
- ○ D 81

One Man Band

 Rory Blackwell, a man from the United Kingdom, once managed to play more than 100 instruments at the same time all by himself. The instruments included a double left-footed perpendicular percussion pounder and a 12-outlet horn blower. It makes you wonder if everyone in the audience was wearing earplugs!

3. Andrea learned how to play the oboe, the tuba, and the drums all at once. She performed several concerts in her backyard for her friends. She charged each of her friends the same admission fee. The table below shows the total amount of admission money she collected for her Thursday, Friday, and Saturday shows.

Concert Admission Fees

Show	Number of Friends	Amount of Money Collected
Thursday	9	$27.00
Friday	12	$36.00
Saturday	6	$18.00
Sunday	7	?

On Sunday, 7 of Andrea's friends came to her show. How much money did Andrea collect in admission fees on Sunday?

- ○ A $21.00
- ○ B $28.00
- ○ C $48.00
- ○ D $81.00

Subject Review

A **pattern** is a series of numbers or shapes that follows a rule.

Below is a pattern of numbers. The pattern shows each number increasing by 2.

1, 3, 5, 7, 9, 11

Here is a pattern of shapes. The pattern shows two squares and then two circles repeating.

Some math problems require you to make predictions about patterns. To figure out number patterns, you usually need to do some calculations. To figure out shape patterns, you'll probably need to count the shapes.

Remember those questions on page 61? Here are the answers.

What does Maurice Bennett do with his burnt toast?
Maurice Bennett, the Toastman, is an artist from New Zealand. He creates pieces of art out of burnt toast. He even made a giant portrait of Elvis Presley with toast!

What kind of weird machine did Martin Goetze invent?
Martin Goetze invented a machine that makes dimples. Strange, huh?

How many instruments is it humanly possible to play at once?
Rory Blackwell once played more than 100 instruments all at the same time. It might be possible to play more, but you probably wouldn't want to hear it!

CHAPTER 9
Variables

What in the world are "moon trees"?

How did ancient Hebrews do math if they didn't have any numbers?

Who discovered Coney Island, the home of the famous roller coaster, the Cyclone?

Variables

A **variable** is a symbol that represents a number. A variable might be an empty box, a letter, or a symbol. Look at the problem below.

$$2 + \square = 8$$

The box in this problem stands for a number, but you don't know what it is. Your job is to figure out what number goes in the box to make the problem correct. Read the problem again. It asks you, "2 plus *what* equals 8?" Well, count up. 8 is 6 more than 2, so 6 belongs in the blank. Check your work. $2 + 6 = 8$. Good job! You can also find the answer by subtracting. $8 - 2 = 6$.

Try another example with a letter instead of a box.

$$5 \times n = 20$$

Read it this way: "5 times *what* equals 20?" You can guess and check. Does 5 times 1 equal 20? No. Does 5 times 2 equal 20? No. Does 5 times 3 equal 20? No. Does 5 times 4 equal 20? Yes! The variable *n* equals 4.

You can also try to work backward to find the variable. You know that 5 times something equals 20, so divide 20 by 5 to get 4. That's the same answer as before.

Sometimes a multiplication problem with a variable in it will be written like "$2n = 14$." The $2n$ is just another way of writing $2 \times n$. Solve this one. 2 times *what* equals 14? If you said 7, you're absolutely correct!

Traveling Trees

In 1971, the *Apollo 14* command module was launched into space. One astronaut, Stuart Roosa, brought some tree seeds from Earth with him on the trip around the moon. He wanted to learn if the seeds could still grow after being in a spaceship with very little gravity. When *Apollo 14* came back to Earth, Roosa gave his seeds to the United States Forest Service. The Forest Service planted the seeds, and in a short time they sprouted into tiny trees.

The Forest Service grew hundreds of these trees. They were nicknamed "moon trees." So many of the moon trees grew that the Forest Service didn't have enough room to keep them. The Forest Service gave them away to people all over the country. Who knows? You might have a moon tree right in your backyard!

Directions: Answer the question based on the passage Traveling Trees.

▶ Javier collected seeds from a moon tree to give to his best friends. Javier wants to distribute the seeds evenly between his friends. He wrote down the following math problem to help him figure out how many seeds he should give to each of his friends.

$$9x = 45$$

Find the value of x in the problem.

○ A 36
○ B 455
○ C 54
○ D 5

Know It All Approach

Read the problem carefully. You need to figure out what the variable x stands for. Read the problem like this: "9 times *what* equals 45?" You can guess and check or divide. Because $45 \div 9 = 5$, $9 \times 5 = 45$. So x must equal 5.

Retrace your steps and answer your question a second time. Make sure the numbers you used in your calculations match the ones in the problem. Make sure you calculated correctly. Then find the answer choice that agrees with the answer. Fill in the bubble next to answer choice (D).

A strategy you can use when you don't know which answer choice is correct is to get rid of the wrong answer choices first. Think, "9 times *what* equals 45?" You know that 9 must be multiplied by some number to get 45. Any answer choice greater than 45 is wrong. Get rid of answer choices (B) and (C). Now you are left with two choices to decide between.

Alphabet Math

The ancient Hebrews didn't have numbers like the kind we use today. Instead, they used letters from the Hebrew alphabet in place of numbers.

1. According to the table, which of the following number sentences is true?

Hebrew Letter	א	ב	ג	ד	ה	ו	ז	ח	ט
Name	Aleph	Bet	Gimel	Dalet	Hey	Vav	Zayin	Chet	Tet
Number	1	2	3	4	5	6	7	8	9

○ A 6 − ב = 8
○ B 6 + ב < 7
○ C 6 − ב > 7
○ D 6 + ב = 8

2. Use the chart of Hebrew numerals to help you solve the problem below.

$$2 \times ט = ?$$

What is the answer?

○ A 18
○ B 16
○ C 14
○ D 12

On the Boardwalk

 Henry Hudson was the first European to discover Coney Island, in New York. That was in 1609. In the 1800s, it became a popular vacation spot for the wealthy. In 1884, Coney Island became the home of America's first modern roller coaster. Soon afterward, more amusement park rides and other attractions were added, including the famous Cyclone.

Coney Island is still very popular with tourists today. During the summer, visiting crowds purchase roughly 50,000 hot dogs and 20,000 orders of fries every weekend!

3. Sadie and her friends went to Coney Island. Sadie bought 2 hot dogs for each of her 5 friends. She then dropped some hotdogs and had seven hot dogs left. How many hot dogs did Sadie drop?

$$2 \times 5 - y = 7$$

Find the value of y in this problem.

- ○ A 2
- ○ B 3
- ○ C 7
- ○ D 9

4. Sadie ordered fries for all of her friends. The man at the counter had only 80 fries to divide among Sadie's friends. Sadie used this problem to help the man figure out how many fries to put into each bag.

$$5x = 80$$

Find the value of x in this problem.

- ○ A 3
- ○ B 15
- ○ C 16
- ○ D 75

Subject Review

A **variable** is a symbol that represents a number. If a math problem has a variable in it, you probably have to figure out the value of the variable.

A variable in a math problem might be a box, a letter, or a symbol. Here are some examples of problems with variables.

$5 - \Box + 1 = 2$

$4 + d = 8$

$8x = 32$

Now the answers to the earlier questions are below.

What in the world are "moon trees"?

Moon trees are trees grown from seeds that astronaut Stuart Roosa took with him on his journey around the moon. You can find one of the trees at the White House.

How did ancient Hebrews do math if they didn't have any numbers?

Ancient Hebrews used letters of their alphabet to represent numbers. Think about it: Our numbers are made up of symbols, too. Try using our ABCs to represent numbers.

Who discovered Coney Island?

Dutch explorer Henry Hudson discovered Coney Island in 1609. The famous Hudson River was named after him too!

Know It All! Elementary School Math

Directions: Read the passages below and answer the questions that follow.

Cool Cow

 What do cows do when they're not acting on a movie set? They chill out in their hotel rooms, of course! A cow that appeared in a recent Jim Carrey movie stayed in a fancy hotel in New York City. The hotel turned its ballroom into a cozy living space for the cow. Other luxury hotels have rolled out the red carpet for different kinds of animals as well.

1. Imagine three cows sharing a hotel room and ordering from room service. The figures below show how much of a sandwich each cow ate. What can you decide from these pictures?

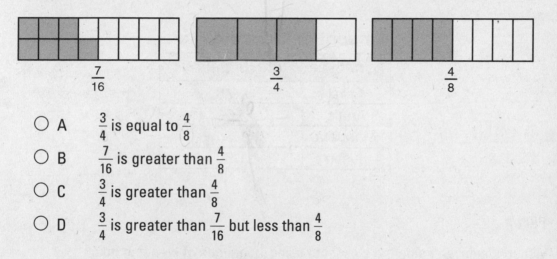

$\frac{7}{16}$ $\frac{3}{4}$ $\frac{4}{8}$

- ○ A $\frac{3}{4}$ is equal to $\frac{4}{8}$
- ○ B $\frac{7}{16}$ is greater than $\frac{4}{8}$
- ○ C $\frac{3}{4}$ is greater than $\frac{4}{8}$
- ○ D $\frac{3}{4}$ is greater than $\frac{7}{16}$ but less than $\frac{4}{8}$

2. A hotel manager asked her staff which animals made the best hotel guests. The chart shows the results of her survey.

Best Hotel Guest	Number of Votes
Cow	3
Cat	13
Dog	20
Spider	4
Total	**40**

What percent of the staff members liked dogs the best?

Answer: _____

3. Mike has 96 pounds of straw to make beds for the 3 cows. He used this number sentence to figure out how many pounds of straw he should use for each bed.

$$3n = 96$$

Find the value of n in the problem.

○ A 32
○ B 99
○ C 288
○ D 93

4. Four employees cleaned out the ballroom after the cows left. The chart below shows how much cleaning fluid each employee used to scrub out the room.

Amount of Cleaning Fluid

Employee	Fluid (in gallons)
Mike	0.75
Annie	1.60
Manuela	1.55
Justin	1.30

Part A

Which employee came the closest to using $1\frac{1}{2}$ gallons of cleaning fluid?

Answer: _____

Part B

Which employee used more cleaning fluid than Mike and less than Manuela?

Answer: _____

And You Thought Climbing Was for Monkeys

 Bet you didn't know there was a world championship pole-climbing competition! The competition takes place in Hampshire County, England. In 1999, Jeremy Barrell shimmied up an 80-foot pole in just 10.75 seconds!

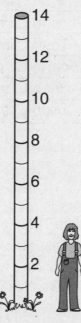

5. Teddy has been practicing pole climbing. Teddy climbs 2 feet higher each time he climbs the pole. If Teddy climbed 8 feet on his last try, how many more climbs will it take for him to reach 14 feet? Explain or show how to find your answer.

_____

_____

_____

_____

6. Teddy's older sister Georgia decided to try climbing the pole, too. Teddy wrote down how many feet Georgia climbed on each of her 5 attempts. If Georgia's next climb follows the pattern shown, how many feet will Georgia probably climb on her sixth try?

Georgia	
First Climb:	7 feet
Second Climb:	9 feet
Third Climb:	13 feet
Fourth Climb:	19 feet
Fifth Climb:	27 feet
Sixth Climb:	?

- ○ A 29
- ○ B 31
- ○ C 35
- ○ D 37

7. Teddy decided to enter a local pole-climbing competition. The entry fee was $40. He had already saved $13. He used this problem to find out how much more money he needed to save. Help Teddy find the value of y.

$$13 + y = 40$$

- ○ A $27
- ○ B $53
- ○ C $26
- ○ D $13

CHAPTER 10
Two-Dimensional Figures

How did Alexandra Nechita earn her nickname, "the petite Picasso"?

How far can a person travel on a lawn mower?

Why can't astronauts eat bread? What do they eat instead?

Do you like building things? Perhaps you plan to grow up to be a construction worker, an engineer, a scientist, an inventor, a sculptor, or an architect. Even if you don't want to do any of those things, two-dimensional figures will show up throughout your everyday life.

Two-Dimensional Figures

A **two-dimensional figure** is a flat figure that has length and width.

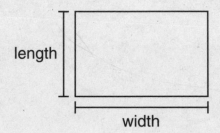

Figures are made of **line segments.** A **line** goes on forever in two directions. **Line segments** have endpoints. Some figures have **parallel** line segments. Parallel lines are lines or line segments that never cross. In the rectangle above, there are two sets of parallel line segments.

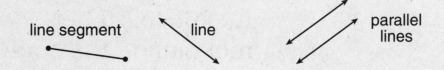

Figures can also contain rays. A **ray** has an endpoint and goes on forever in one direction. When two rays share the same endpoint, they form an **angle.** A **right angle** is a special type of angle that forms square corners.

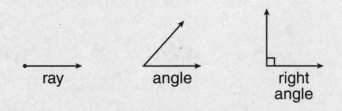

A **polygon** is a closed two-dimensional figure made entirely of straight sides. An open shape is not a polygon. A shape with curved sides is not a polygon. Compare the following two groups:

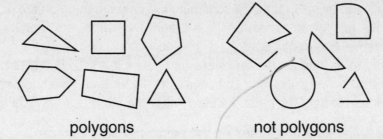

polygons   not polygons

There are many different types of polygons. Polygons are named according to the number of sides. Take a look at the chart below.

Polygon Name	Number of Sides	Number of Angles	Special Features
Triangle	3	3	A triangle has no parallel sides.
Square	4	4	A square has four right angles and two pairs of parallel sides. The sides are all equal in length.
Rectangle	4	4	A rectangle has four right angles and two pairs of parallel sides.
Trapezoid	4	4	A trapezoid has only one pair of parallel sides.
Pentagon	5	5	
Hexagon	6	6	

Petite Picasso

Have you ever dreamed of becoming the next big thing on the art scene? That's just what happened to Alexandra Nechita, a young artist from Romania. She began drawing when she was just two years old. She began painting at four years. Alexandra had her first art exhibit in Los Angeles when she was only eight! Critics around the world praised her work. They nicknamed her "the petite Picasso" after the famous artist Pablo Picasso.

Directions: Answer the question based on the passage Petite Picasso.

▶ Tyrell wants to be a famous artist someday like Alexandra Nechita. Instead of painting, Tyrell likes to make designs with different pieces of paper. Tyrell made one design using 1 rectangle and 2 triangles. Which of the following could be that design?

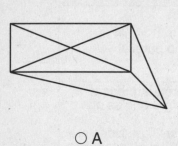 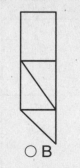  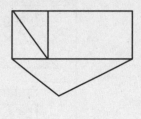

○ A   ○ B   ○ C   ○ D

Know It All Approach

Read the question carefully. This question asks you to identify the answer choice that has certain shapes. The important words and numbers are *1 rectangle and 2 triangles.*

To answer the question, look at the designs in each answer choice. Count the number of triangles and rectangles in each answer choice and write down your totals under each drawing. Make sure to count every shape in each design. There are 6 triangles in answer choice (A). There are 3 triangles and 1 rectangle in answer choice (B). There are 2 triangles and 1 rectangle in answer choice (C). There are 3 triangles and 1 rectangle in answer choice (D).

Count the shapes in each design again. Make sure you didn't forget to count any shapes or count some shapes twice. Only answer choice (C) contains 1 rectangle and 2 triangles. Fill in the bubble next to answer choice (C).

If you were not sure of the correct answer, you could get rid of answer choices that have more than 2 triangles.

Directions: Read the passages below and answer the questions that follow.

Traveling Man

Gary Hatter from Champaign, Illinois, drove more then 14,000 miles on his lawn mower. He traveled through 48 states and parts of Mexico and Canada as well. Hatter did this to raise money for surgery that he needed. Along his amazing 260-day journey, he received enough donations to cover the cost of the operation, and he set a new world record!

1. After reading about Gary Hatter, Mr. Leoni noticed that he needed to mow his lawn. While he was cutting the grass, Mr. Leoni decided to mow some fun shapes into his lawn.

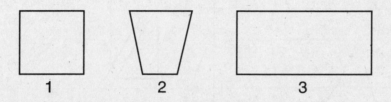

How is shape 1 more like shape 3 than shape 2?

○ A Shapes 1 and 3 have sides that are all the same size.
○ B Shapes 1 and 3 have right angles.
○ C Shapes 1 and 3 do not have right angles.
○ D Shapes 1 and 3 have no parallel lines.

2. Mr. Leoni mowed a pentagon, a hexagon, and then a trapezoid in that order. Which of the following shows how Mr. Leoni's lawn looked when he was finished?

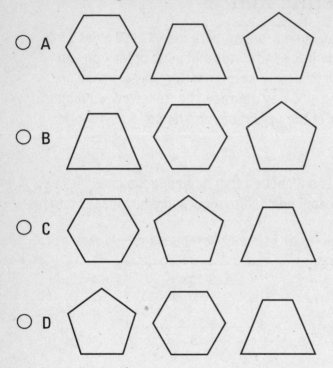

○ A

○ B

○ C

○ D

No Bread Onboard

Did you ever wonder what it's like to eat in space? The lack of gravity causes all sorts of weird mealtime problems. For example, astronauts in the space shuttle can't eat bread because it has a lot of crumbs. The crumbs would float all over the space shuttle and make a big mess. Instead, astronauts eat a less "crumby" type of bread—tortillas. Maybe the round shape of the tortillas reminds the astronauts of Earth while they are traveling.

3. Melinda and Rose wanted to know if tortillas really didn't make many crumbs. To find out, they took some tortillas and ripped them up into the shapes below.

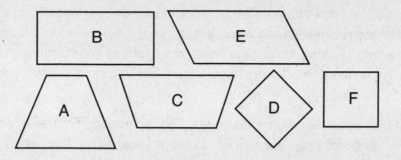

When they were done, Melinda gathered all the tortillas that were like squares and rectangles, and Rose gathered up all the tortillas that were like trapezoids.

Write the letters of the tortillas in the boxes below to sort them into groups.

Melinda's Group

Rose's Group

Describe the special features shared by the tortillas in each group.

Melinda's Group: _____

_____

_____

_____

Rose's Group: _____

_____

_____

Subject Review

In this chapter, you learned about different kinds of two-dimensional figures. Now, can you see these figures everywhere in the world around you? There are lines hiding out in your shoelaces, rectangles lurking in your toasters, pentagons in your birdhouses, and other figures to find. Think about the figures in your life while you review what you've learned in this chapter.

A **two-dimensional figure** is a figure that has length and width. Figures are often made up of line segments. **Parallel lines** are two lines that run next to each other without ever crossing. You can describe some shapes by whether or not they have parallel lines.

Figures can also be made with rays. **Rays** have one endpoint and the other end goes on forever. When two rays meet, they form an **angle.** Special angles with square corners are **right angles.**

A **polygon** is a closed two-dimensional figure made entirely of straight sides. Polygons get their names from the numbers of sides they each have.

Here are the answers to the questions from the beginning of the chapter.

How did Alexandra Nechita earn her nickname, "the petite Picasso"?

Alexandra Nechita began painting at the age of two. She was such a wonderful young artist that people compared her to Pablo Picasso, a very famous painter.

How far can a person travel on a lawn mower?

Gary Hatter drove through 48 states and parts of Mexico and Canada on his lawn mower. In total, he covered more than 14,000 miles. Perhaps he even mowed some lawns along the way.

Why can't astronauts eat bread? What do they eat instead?

Bread leaves crumbs can seriously mess up the space shuttle. Instead, astronauts eat tortillas because these flat, round breads don't leave as many crumbs.

CHAPTER 11
Three-Dimensional Figures

What does plastics-factory owner David Morgan do for fun?

Who came up with the idea for the cardboard box?

What was so special about a drum played in Dublin, Ireland's 1999 Saint Patrick's Day Parade?

Get ready to venture into another dimension! In this chapter, you will learn all about three-dimensional figures.

Three-Dimensional Figures

A **three-dimensional figure** has length, width, and height.

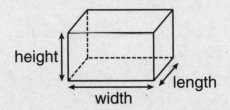

Three-dimensional figures are made up of faces and edges. **Faces** are the sides of a figure, and **edges** are where two faces meet. For example, the front cover of this book is a face, and where the front of the book and the side of the book meet is an edge.

edge
(where two flat surfaces meet)

face
(flat surface)

The chart below shows some different types of three-dimensional figures.

Shape	Number of Faces	Number of Edges	Shapes of the Faces
Cube 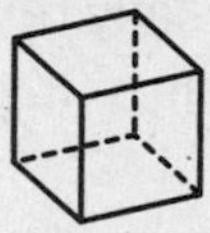	6	12	6 square faces
Rectangular prism 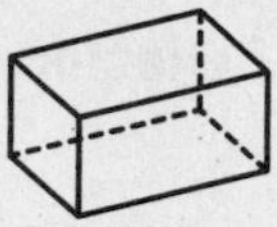	6	12	6 rectangular faces
Triangular pyramid 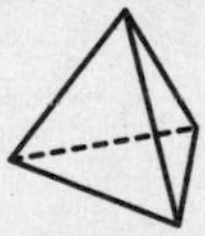	4	6	4 triangular faces
Rectangular pyramid 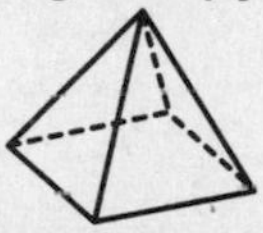	5	8	1 rectangular face 4 triangular faces
Cylinder 	4	0	2 circular faces
Cone 	1	0	1 circular face

Traffic Cone Collector

David Morgan of Oxfordshire, England, has an unusual hobby. He collects traffic cones—you know, those party hat-shaped things you see sometimes on the road. He has a collection of more than 500 of these brightly colored cones. In fact, his whole life practically revolves around traffic cones. That's because he owns a plastics business that produces more than 1 million traffic cones per week.

Directions: Answer the question based on the passage Traffic Cone Collector.

▶ In the space below, draw a cone. Remember that cones are three-dimensional figures. So make sure your drawing shows how the cone is three-dimensional.

Know It All Approach

This question asks you to draw a cone. In the passage about David Morgan, "party hat-shaped things" gives you a little clue about the shape of cones. Before you draw the cone, picture a cone in your mind. You might want to imagine a traffic cone or a party hat to help you. If you need to, look back at the chart of three-dimensional figures on page 87 to help you remember what a cone looks like. Then neatly draw a cone in the space provided.

Double-check your drawing and make sure the cone you drew is three-dimensional. That means that it should look like it has length, width, and height. As long as you drew a cone that looks three-dimensional, then you have completely answered the question. Make sure that you drew the cone clearly.

Directions: Read the passages below and answer the questions that follow.

Paper That Helps You Move

A box made of paper? Why, that's madness! That's what some people probably said to Robert Gair in 1890 when he invented the cardboard box. Before then, boxes were often made out of wood. Today, millions of people around the world rely on using Gair's invention to ship and store items. Just imagine what moving to a new house would be like if we didn't have cardboard boxes!

1. Gina and her mother made some three-dimensional figures by cutting pieces of cardboard from an old box. The shapes are shown below.

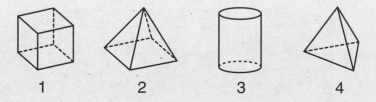

Which figure has five faces and eight edges?

- ○ A 1
- ○ B 2
- ○ C 3
- ○ D 4

2. Gina and her mother also made the figures below.

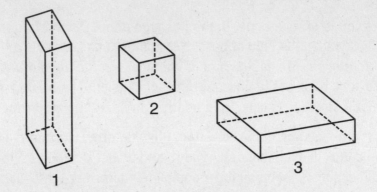

What do all three of the figures have in common?

- ○ A They are all cubes.
- ○ B They each have five faces.
- ○ C They are all rectangular pyramids.
- ○ D They each have twelve edges.

A Drum You Can Stand In

In 1999, a special drum was played at the Saint Patrick's Day Parade in Dublin, Ireland. Of course, most parades have drums, but this drum was different. It measured about 15.5 feet across and was more than 6 feet deep. At the time, it was the world's largest drum.

3. Look at the picture of the drum below.

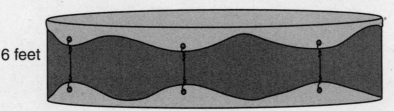

6 feet

15 feet

What three-dimensional shape best describes the drum?

Answer: _____

4. Mr. Martin, the music teacher, asked his students to design their own musical instruments. One student made a cube-shaped drum. In the space below, draw a cube.

How many faces does the drum have?

Answer: _____

How many edges does the drum have?

Answer: _____

Subject Review

It's time to make a drum out of a traffic cone and put it in a box! No . . . it's time to review what you've learned in this chapter.

A **three-dimensional figure** has length, width, and height. Three-dimensional figures are made up of **faces** and **edges. Faces** are the flat sides of a figure, and **edges** are where two faces meet.

Some three-dimensional figures are

- cubes
- rectangular prisms
- triangular pyramids
- rectangular pyramids
- cylinders
- cones

Remember the questions from the beginning of the chapter? Here are the answers.

What does plastics-factory owner David Morgan do for fun?
It may sound odd, but David Morgan collects traffic cones.

Who came up with the idea for the cardboard box?
Robert Gair invented the cardboard box in 1890. Who knows—maybe the next box-making idea will come from you!

What was so special about a drum played in Dublin, Ireland's 1999 Saint Patrick's Day Parade?
It was enormous. This drum measured about 15.5 feet wide and more than 6 feet deep.

CHAPTER 12
Grids

What remarkable feat did Walter Port and Eric Sevareid perform in 1930?

How did six Colorado teenagers get to spend eighteen days in the Alaskan wilderness?

Where is the most famous hedge maze in the world located?

In this chapter, you will learn how to use grids. Grids are used to show all sorts of information in math.

Plotting Points on a Grid

A **coordinate grid** is made up of horizontal and vertical lines that cross one another. It looks like this.

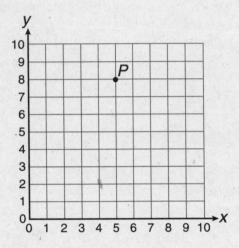

The number line at the bottom of the grid is called the **x-axis.** On the x-axis, the numbers become greater in value as you move from left to right.

The number line on the side of the grid is called the **y-axis.** On the y-axis, the numbers become greater in value as you move from the bottom toward the top of the grid.

An **ordered pair** is a pair of numbers that gives a location on a graph, map, or grid. The first number in the ordered pair is on the x-axis. The second number in the ordered pair is on the y-axis.

You can use an ordered pair on a grid to plot a **point.** A point is an exact place on a grid. The ordered pair for point P on the grid above is (5,8). The first number indicates the position of the point along the x-axis. The second number indicates how far up the point is along the y-axis.

You might be given an ordered pair and asked to plot that point on a grid. For example, to plot the point (4,2) on a coordinate grid, locate the 4 on the *x*-axis. Put your finger or your pencil on the number. Then move your finger or pencil up the 4 line until you reach the 2 line of the *y*-axis. Draw a point at the place where the 4 line of the *x*-axis and 2 line of the *y*-axis cross.

Tip

Remember that the first number in an ordered pair shows the location along the *x*-axis. The second number in an ordered pair shows the location along the *y*-axis. The most common mistake is to mix these two numbers up!

Finding Patterns on a Grid

You can follow patterns on a grid to find points. Look at the coordinate grid below. It shows a path that an ant makes while walking across a map. The ant starts by walking three units to the left. Then the ant turns and walks two units down. The ant turns again to walk three units left, two units down, and three units to the left. This ant is walking in a pattern!

If the ant continues walking this pattern, the ant will walk to point B. Put your finger or pencil on the path and see if this is true.

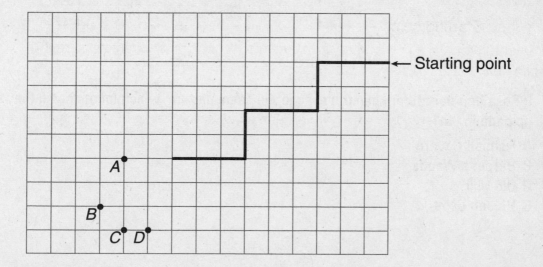

← Starting point

Row, Row, Row Your Boat

In 1930, two teenagers, Walter Port and Eric Sevareid, made a remarkable journey. In their small canvas canoe, they paddled up the Mississippi River from Minneapolis, Minnesota, into the Canadian wilderness. The two boys traveled nearly 2,300 miles in the icy waters by canoe.

Directions: Answer the question based on the passage Row, Row, Row Your Boat.

▶ Inspired by the story of Walter Port and Eric Sevareid, Rosa and her father decided to take a canoe trip along a local river. Their journey is mapped below.

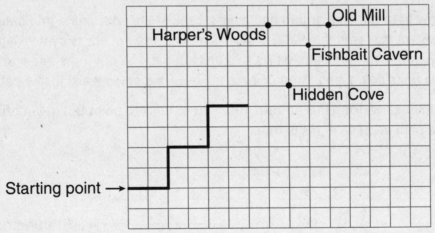

If Rosa and her father keep traveling in the same pattern, which location will they end up paddling to?

○ A Fishbait Cavern
○ B Harper's Woods
○ C Old Mill
○ D Hidden Cove

Know It All Approach

Read the question and study the map carefully. This question asks you to identify a point on a grid map. In this question, the important words are *same pattern* and *which location*.

The grid map shows you the beginning of Rosa and her father's journey. It shows that they traveled 2 squares to the right and 2 squares up, then 2 squares to the right and 2 squares up, and so on. On the grid map, continue that same pattern using your pencil. Draw a line going 2 squares up, 2 squares to the right, 2 squares up, and 2 squares to the right, until the line cannot go any further. If you continued the pattern correctly, the line should end at the point labeled "Old Mill."

Make sure you have read the question correctly and that you have counted the squares correctly. Then, look at the answer choices. Answer choice (C) is Old Mill. Just to be on the safe side, read all of the other answer choices to make sure that you are correct.

Finally, clearly fill in the bubble for answer choice (C).

Directions: Read the passages below and answer the questions that follow.

Teenage Trekkers

In 1996, a group of six teenagers from Colorado were the first ever to win a trip to Alaska through the Adventure Grants program. The teenagers, along with their guides, spent eighteen days backpacking and canoeing through Arctic National Park and Preserve in Alaska. They endured harsh weather and got to see grizzly bears, moose, and many other animals in their natural habitat.

1. George is interested in learning more about Alaskan wildlife. He decided to go see some Alaskan animals at the local zoo.

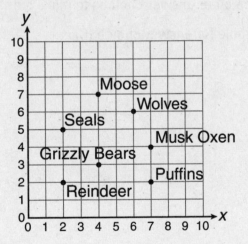

Write an ordered pair that describes the location of the animal closest to the grizzly bears at the zoo.

Answer: _____

The Hedge Maze of Hampton

You've probably found your way through mazes on paper, but imagine what it would feel like if you were actually inside the maze. You could find out for yourself if you ever visited the giant hedge maze at Hampton Court in London, England. A hedge maze is a maze made out of hedges, shrubs, or leafy trees. The world-famous hedge maze in Hampton Court Palace Gardens has been around since 1702. It covers one-third of an acre, and is made of hedges that are taller than most people.

2. Julia is walking on a path surrounded by hedges at her local park. A coordinate grid of the park is shown below.

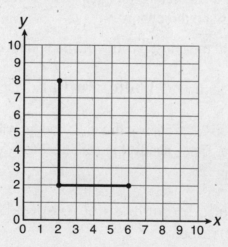

The park is in the shape of a rectangle. Three corners of the rectangle are plotted on the grid. Which ordered pair describes the location of the fourth corner of the rectangle?

- ○ A (8,6)
- ○ B (2,6)
- ○ C (8,2)
- ○ D (6,8)

Subject Review

In this chapter, you learned how to read and plot points on grids. It's time to review what you've learned.

A **coordinate grid** is made up of horizontal and vertical lines.

An **ordered pair** is a pair of numbers that gives a location on a graph, map, or grid. The first number in the ordered pair is on the **x-axis,** which has numbers going from left to right. The second number in the ordered pair is on the **y-axis,** which has numbers going from bottom to top.

You can use an ordered pair on a grid to plot a point. A **point** is an exact place on a grid.

You may find patterns on coordinate grids by looking at the number of units that a path covers. Count the number of units that a path goes in the vertical (up and down) direction and in the horizontal (left to right) direction.

Now, here are the answers to the questions frm the beginning of the chapter.

What remarkable feat did Walter Port and Eric Sevareid perform in 1930?

They paddled their canoe 2,300 miles into the Canadian wilderness from Minneapolis, Minnesota. That's almost the distance from New York, New York, to Los Angeles, California.

How did six Colorado teenagers get to spend eighteen days in the Alaskan wilderness?

They were the first teenagers to win a trip through the Adventure Grants program.

Where is the most famous hedge maze in the world located?

The maze is located in Hampton Court Palace Gardens in London, England.

CHAPTER 13
Transformations and Symmetry

How hot does it get inside
the Sun?

How many number
symbols did the
Babylonians have?

What else can you do with
spaghetti besides eat it?

More shapes are coming your way! This chapter will discuss what happens to shapes when you move them. You'll also learn about the symmetry of shapes.

Transformations

Transformations are ways to move figures. There are three types of transformations: slides, flips, and turns.

A slide moves a figure from one place to another without changing its shape. Think of what a playground slide is like. When you sit on a slide, you move from the top to the bottom. You should still be sitting up at the end. Look at an example of a mathematical slide.

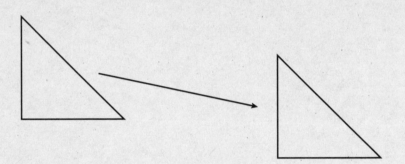

A flip moves a figure over a line. Imagine the pages of a book. When you flip the page, you look at the back of the page. A flip is also called a reflection. You can place a mirror on the dotted line in the example below and see the image that's on the other side of the line flipped.

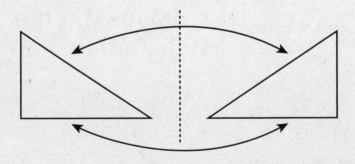

A turn moves a figure around a point. Think of the hands of a clock. They move around the center point of the clock. As each hand moves around the clock, it is in a different position from before. A turn is also called a rotation.

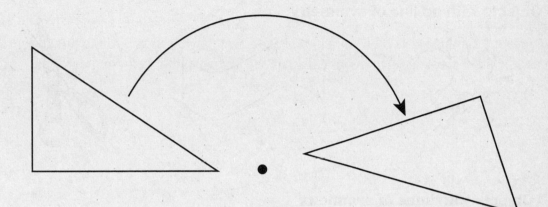

Symmetry

A line of symmetry is a line through an object that divides it into equal parts. Objects that have lines of symmetry are symmetrical. Objects may have more than one line of symmetry. See the examples below.

Objects with no line of symmetry

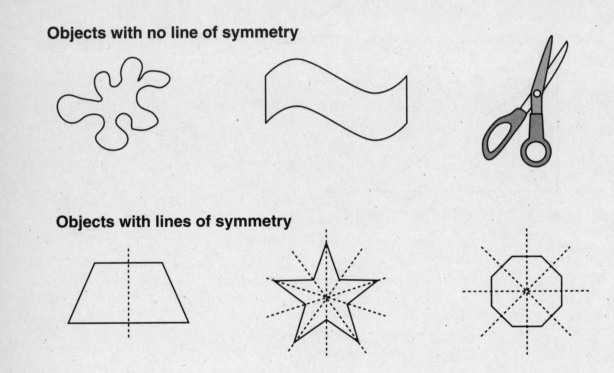

Objects with lines of symmetry

You can find lines of symmetry by folding the figure in half and seeing if the two halves match up exactly. Objects with no lines of symmetry will not fold in half exactly.

Whew! You Think It's Scorching Hot Here!

 You might think the Sun can feel pretty hot down here on Earth, but it may be even hotter than you realize. The surface of the Sun reaches 10,000 degrees Fahrenheit. That's nothing compared to the temperature inside the Sun, though. At the sun's core, temperatures soar to more than 30 million degrees Fahrenheit!

Directions: Answer the question based on the passage Whew! You Think It's Scorching Hot Here!

▶ The drawing below shows a sun that has been cut along a line of symmetry.

How many rays were on the sun before it was cut in half?

○ A 10
○ B 8
○ C 4
○ D 12

Know It All Approach

Read the question carefully and look at the figure. You know that this shape has a line of symmetry and that the other half of the figure was cut off. To answer this question, study the drawing of the sun.

The remaining half-sun has 4 rays. The missing half must also have 4 rays. So the total number of rays must be 8, answer choice (B).

You can try to draw the missing half of the figure to check that this answer is correct. Finally, fill in the answer bubble that goes with your answer choice.

Directions: Read the passages below and answer the questions that follow.

Babylonian Math

We use the base-ten number system. This number system is based on ten numbers, for which we use the symbols 0, 1, 2, 3, 4, 5, 6, 7, 8, and 9. The ancient Babylonian number system was based on 60 numbers. Before you start feeling sorry for those poor Babylonian kids who had to memorize 60 number symbols, listen to this: The Babylonians had only two number symbols! They just positioned those two symbols in different ways to represent the 60 numbers in their system. Those Babylonian kids didn't have it so bad after all!

1. The Babylonians used the symbol below to represent the number 10.

Which figure would result if you used a slide to move the symbol across the dotted line?

      ○ A   ○ B   ○ C   ○ D

Don't Sit Across the Table from Kevin

If you eat lunch in your school's cafeteria, you've probably seen milk come out of someone's nose at least once. But have you ever seen spaghetti fly out of someone's nose? In 1998, Kevin Cole shot a strand of spaghetti out of his nose. The spaghetti strand flew almost 8 inches away from him!

2. Nigel spent weeks practicing how to blow spaghetti out his nose. He got so good at it that he could make the spaghetti strands land in the shape of alphabet letters. Nigel made the letters below. Which letter has more than one line of symmetry?

○ A **D**

○ B **X**

○ C **M**

○ D **V**

Subject Review

A **transformation** is a way to move a figure. There are three types of transformations: slides, flips, and turns.

A **slide** moves a figure along a path without changing its appearance.

A **flip** moves a figure over a line so that the final figure looks like a reflection.

A **turn** moves a figure around a point.

When a shape is folded to create equal parts, this is known as **symmetry.** An object that has symmetry is symmetrical.

The fold line that divides a shape into equal parts is called a line of symmetry. Some shapes have only one line of symmetry. Some shapes have more than one line of symmetry. Other shapes have no line of symmetry.

Remember the questions from page 101? Here are the answers.

How hot does it get inside the Sun?
The temperature at the Sun's core reaches more than 30 million degrees Fahrenheit. Imagine how quickly you could cook a million hot dogs.

How many number symbols did the Babylonians have?
The Babylonians had only two number symbols. But they positioned the symbols in different ways to represent the 60 numbers in their number system.

What else can you do with spaghetti besides eat it?
You can blow it out of your nose like Kevin Cole! Don't sit across from him at the dinner table.

CHAPTER 14

Perimeter and Area

What do Keanu Reeves, Freddie Prinze Jr., and the Backstreet Boys have in common?

How big were prehistoric roaches?

How does the Texas horned lizard scare off its enemies?

In this chapter, you'll learn how to calculate a figure's perimeter and area.

Perimeter

Perimeter is the distance around the edge of a closed figure. To find the perimeter, add the measures of all the sides of a figure together.

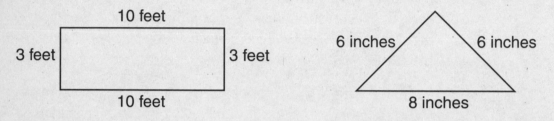

The perimeter of the rectangle is the sum of 10 feet, 3, feet, 10 feet, and 3 feet. 10 + 3 + 10 + 3 = 26 feet.

The perimeter of the triangle is the sum of 6 inches, 6 inches, and 8 inches. 6 + 6 + 8 = 20 inches.

Area

Area is the amount of space inside a closed figure. To find the area of a figure, just count the number of square units inside it.

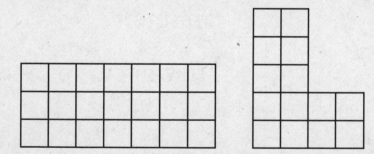

Count the number of units (squares) inside the rectangle. How many units are there? The area of the rectangle is 21 square units. Now count the number of units inside the L-shaped polygon. How many units are there? The area of the L-shaped polygon is 14 square units. Great job!

Watch Where You're Aiming That Thing!

 Paintball is becoming one of the most popular extreme sports around. It's like hide-and-seek, dodgeball, and tag all rolled up into one game. In paintball, groups of players shoot balls of paint at one another. The object is to stay in the game as long as possible without being hit.

Getting pegged by a paintball doesn't exactly tickle. Paintballs fly fast and hit hard, so players must wear lots of protective gear. Paintball players are required to wear special helmets and jumpsuits to avoid injuries (and paint-splattered clothing).

The game has become so popular that many celebrities have caught paintball fever. Actors Keanu Reeves and Freddie Prinze Jr., pop idols the Backstreet Boys, and even members of the British royal family have tried their hand at paintball in recent years.

Directions: Answer the question based on the passage Watch Where You're Aiming That Thing!

▶ Kelly is practicing shooting paintballs at this square target. The picture below shows the size of the target.

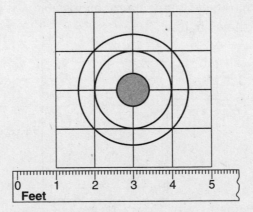

What is the area, in square feet, of the entire square target shown above?

Answer: _____

Know It All Approach

Read the question carefully. This short-response question asks you to figure out the area of the target in square feet. Read the question again to make sure you understand what it's asking.

Now study the diagram. The diagram shows the target with a measuring tape underneath it. The target is divided up into squares, or units. To find the area of an object, you need to count the number of units inside it. How many units are inside the target? If you said 16, you're correct.

Look at the measuring tape. Notice that it is labeled "Feet." Now compare the measuring tape to the target. Each square unit inside the target is exactly 1 square foot. You counted 16 units inside the target, so the area of the target is 16 square feet.

Double-check your answer just to make sure you counted correctly and that you read the measuring tape correctly. Then neatly write your answer, "16 square feet," on the line.

Directions: Read the passages below and answer the questions that follow.

Attack of the Giant Roaches!

Scientists have learned that the state of Ohio was once just a big swamp full of roaches. Of course, that was a few hundred million years ago.

In 2001, scientists in Ohio dug up a complete cockroach fossil that was 3.5 inches long. That's about twice the size of the modern American roach!

1. A group of scientists carefully chipped a roach fossil out of a stone. The piece of rock that held the fossil is shown below.

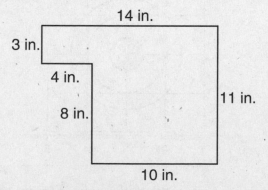

What is the perimeter, in inches, of the rock?

Answer: _____

Know It All! Elementary School Math

Don't Mess with the Texas Horned Lizard

The Texas horned lizard does many things to avoid being gobbled up by its enemies. Its natural coloring helps it to blend in with the environment. It also puffs its body up and hisses at enemies. But wait—that's not all. When the Texas horned lizard gets really freaked out, it squirts blood from little holes near its eyes. That's enough to gross out even the fiercest of wild animals!

2. Below is a drawing of a Texas wildlife exhibit at the zoo.

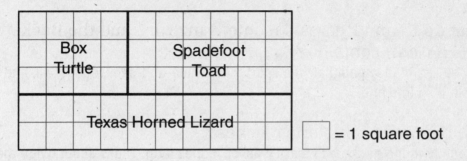

= 1 square foot

What is the area of the cage that holds the Texas horned hizard?

○ A 8 square feet
○ B 11 square feet
○ C 18 square feet
○ D 20 square feet

Subject Review

In this chapter you learned how to figure out the **area** and **perimeter** of shapes. Review what you've learned.

Perimeter is the distance around a closed figure. To find the perimeter of a shape, just add the measures of all the sides together.

Area is the amount of space inside a closed figure. To find the area of a shape, just count the number of square units inside it.

And now, here are the answers to the questions from **page 109**.

What do Keanu Reeves, Freddie Prinze Jr., and the Backstreet Boys have in common?

They all like to play paintball, the sport that is hide-and-seek, dodgeball, and tag all rolled up into one.

How big were prehistoric roaches?

A recently found cockroach fossil was 3.5 inches long. That's about twice the size of roaches today.

How does the Texas horned lizard scare off its enemies?

It changes colors, puffs its body up, and hisses. If those methods don't work, the Texas horned lizard then squirts blood from holes near its eyes.

Directions: Read the passages below and answer the questions that follow.

Hello? Who's Tapping, Please?

In the 1830s, Samuel Morse completely changed the way we communicate over long distances. Morse invented the telegraph, a machine that sends electronic messages in code over a wire. He invented the code, too. His Morse code is a series of dots and dashes that stand for letters and numbers. It can be tapped electronically over a wire, or it can be written down. Even though we have cell phones and computers today, Morse code is still in use across the globe.

1. Omar made up a code of his own. One of the symbols from Omar's code is shown below.

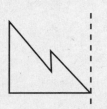

Which figure would result if Omar used a flip to move the symbol across the dotted line?

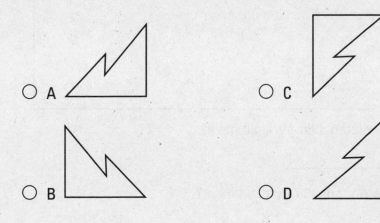

○ A

○ B

○ C

○ D

2. Omar created some words using symbols from his code. Which one of Omar's code words is made of exactly 1 square and 3 triangles?

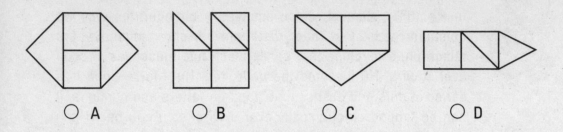

○ A   ○ B   ○ C   ○ D

3. Omar wanted to keep his code a secret. He built a special container in which to hide his secret code. The container was in the shape of a rectangular pyramid. Draw a rectangular pyramid in the space below.

How many faces does a rectangular pyramid have?

Answer: _____

How many edges does a rectangular pyramid have?

Answer: _____

Mush! Mush!

Rachel Scdoris lets nothing stand in the way of her goals. Although she is legally blind, she is quickly becoming one of the world's best sled dog racers. Sled dog racers are called mushers. Rachel first began mushing when she was just 3 years old. When she was 15, she competed in a 500-mile race. At that time, she was the youngest person ever to compete in a sled dog race of that length!

4. Chiyo went sledding around her neighborhood. Her sledding route is mapped below.

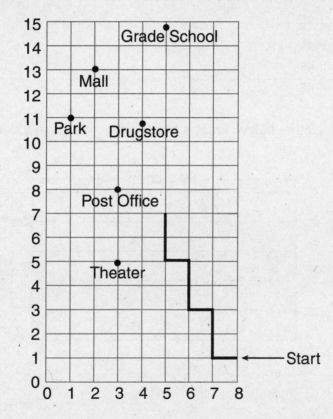

If Chiyo keeps sledding in the same pattern, which location will she sled directly to?

Answer: _____

Write an ordered pair that describes the exact location of the theater on the grid.

Answer: _____

5. Look at the picture of Chiyo's sled below. What is the area of the sled in square units?

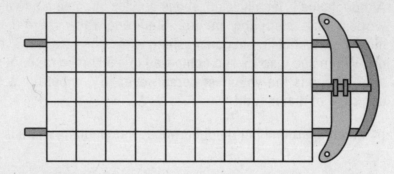

- ○ A 14 square units
- ○ B 36 square units
- ○ C 45 square units
- ○ D 54 square units

6. Based on the picture below, what is the perimeter of Chiyo's sled in inches?

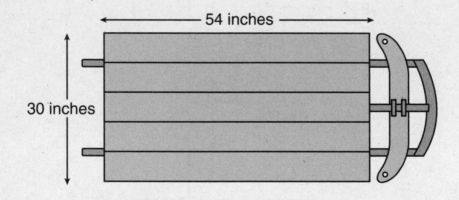

Answer: _____ inches

CHAPTER 15

Measuring Objects and Converting Units of Measurement

What is the largest frog in the world?

How tall can a house of cards get?

What does Mark Hogg like to snack on?

In this chapter, you will learn how to measure objects. You will also learn about different units of measurement.

Using a Ruler to Measure

You can use a ruler to find the length of an object in inches. What is the length of this crayon?

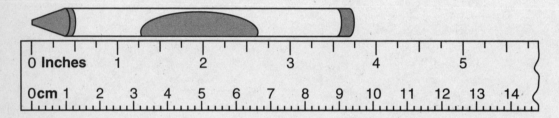

When you measure an object, always make sure that one end of the item is lined up exactly with the zero tick mark on the ruler. Next, count how many whole inches the item is. The crayon is three whole inches and part of another inch. Notice how each inch in broken up into fourths. The crayon covers three of the four parts, so the crayon measures $3\frac{3}{4}$ inches long.

Keep in mind that rulers often have two sides. One side shows inches and the other side shows centimeters.

Units of Measurement

Objects are measured in units. The units of measurement we use are either from the U.S. customary system or from the metric system. The units in the U.S. customary system are listed below.

Length			Weight			Liquid volume		
12 inches	=	1 foot	16 ounces	=	1 pound	8 ounces	=	1 cup
3 feet	=	1 yard	2,000 pounds	=	1 ton	2 cups	=	1 pint
1,760 yards	=	1 mile				2 pints	=	1 quart
						4 quarts	=	1 gallon

Here are the units in the metric system. Notice that the units are based on the number 10.

Length

10 millimeters	=	1 centimeter
10 centimeters	=	1 decimeter
10 decimeters	=	1 meter
1,000 meters	=	1 kilometer

Mass

| 1,000 milligrams | = 1 gram |
| 1,000 grams | = 1 kilogram |

Liquid volume

| 1,000 milliliters | = 1 liter |

Converting Units of Measurement

You can rename units of measurement to make comparisons between them. This is called converting measurements. Try to fill in the blank below.

___ inches = 1 yard

You need to do some calculations to answer this question. You know that there are 12 inches in 1 foot. So, how many inches are in 1 yard? Look back at the U.S. customary system chart. It says that there are 3 feet in 1 yard. To find out how many inches are in 1 yard, you need to multiply 12 inches by 3 feet. You'll find that there are 36 inches in 1 yard.

Now try a problem that uses the metric system.

___ centimeters = 1 meter

According to the chart, there are 10 decimeters in 1 meter and 10 centimeters in 1 decimeter. To find out how many centimeters in 1 meter, you need to multiply 10 decimeters by 10 centimeters. You should find that there are 100 centimeters in 1 meter.

The list below shows how the U.S. customary system and the metric system are related.

- 1 inch is about 2.5 centimeters long.
- 1 foot is about 30 centimeters long.
- 1 yard is a little less than 1 meter long.
- 1 quart is roughly 1 liter.
- 1 ton is roughly 1000 kilograms.

Giant Frog Alert!

The Goliath frog is the largest known frog. Its entire body can measure more than 3 feet long from its feet to its nose. Goliath frogs can weigh 7 pounds or more. That's almost as heavy as a bowling ball!

Directions: Answer the question based on the passage Giant Frog Alert!

▶ Several scientists were studying a group of Goliath frogs. They measured and recorded the length of each frog. The chart below shows the data they gathered.

Name of Frog	Length of Frog
Otis	35 inches
Carla	1 yard
Gertie	$2\frac{1}{4}$ feet
Larry	2 feet 6 inches

```
12 inches = 1 foot
3 feet = 1 yard
```

Which was the longest frog the scientists measured?

○ A Otis
○ B Carla
○ C Gertie
○ D Larry

Know It All Approach

Read the question carefully. It is a multiple-choice question that asks you to compare the lengths of several frogs. Study the chart and the box next to the chart.

The lengths of the frogs are in different units, so you have to convert them to the same unit. The first frog, Otis, is 35 inches. The second frog, Carla, is 1 yard. You have to multiply 12 by 3 to get the number of inches. Carla measures 36 inches. Gertie is $2\frac{1}{4}$ feet. There are 12 inches in each foot, so multiply 12 by $2\frac{1}{4}$ to get 27 inches. Finally, Larry is 2 feet 6 inches. Multiply 2 by 12 to get 24 inches and add 6 to get 30 inches. Larry is 30 inches. Choose the frog with the longest length. Carla is the longest at 36 inches or 1 yard, so answer choice (B) is correct.

Directions: Read the passages below and answer the questions that follow.

Whatever You Do, Don't Sneeze!

 In 1999 Bryan Berg built an enormous house of cards. The house stood more than 7 meters high and used almost 92,000 playing cards!

1. Frank drew a diagram of a house of cards that he wanted to build. Use your ruler to measure the sides of Frank's diagram. Write the missing measurements in the spaces provided.

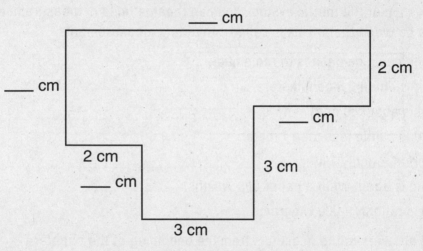

The Worms Crawl In

 In 2000, a man named Mark Hogg once ate 94 worms on a national television show. It took him only 30 seconds to slurp the slippery creatures down. Anyone hungry?

2. When Sandra was working in her garden, she found a worm that was 2 decimeters long. Which of these measurements is the same as 2 decimeters?

○ A 20 millimeters
○ B 20 centimeters
○ C 20 meters
○ D 20 kilometers

Subject Review

In this chapter, you learned about using and converting units of measurement. Review what you've learned.

Rulers often have two sides. One side shows inches and the other side shows centimeters. When you measure an object, always make sure one end of the object is exactly lined up with the zero tick mark on the ruler. If the object measures more than a whole unit, then look at the divisions on the ruler and count the number of covered parts.

Objects are measured in units. The units of measurement we use are from either the U.S. customary system or the metric system. You can rename units of measurement to make comparisons between them. This is called converting measurements.

Here are some basic guidelines to remember.

- 1 inch is about 2.5 centimeters.
- 1 foot is about 30 centimeters.
- 1 yard is a little less than 1 meter.
- 1 quart is roughly 1 liter.
- 1 gram is about what a paper clip weighs.
- 1 ton is roughly 1,000 kilograms.

Here are the answers to the questions from the beginning of the chapter.

What is the largest frog in the world?

The Goliath frog is the largest frog in the world. It can grow to more than 3 feet long and weigh more than 7 pounds.

How tall can a house of cards get?

Bryan Berg built a house of cards that was more than 7 meters high. That's about as high as a real two-story house.

What does Mark Hogg like to snack on?

Worms! Mark Hogg once ate 94 worms in 30 seconds. How gross is that?

CHAPTER 16

Choosing Units of Measurement and Estimating Measurements

Who first came up with the idea for the zipper?

What bizarre talent does Stevie Starr have?

What theater group uses garbage cans and brooms in its performances?

In this chapter, you will learn how to choose the right unit of measurement for the object you are measuring. You will also learn how to estimate (make a guess at) measurements.

Choosing the Right Unit of Measurement

Certain units of measure are better to use in some situations than in others. If you want to measure the distance from your home to your school, it would probably be best to measure the distance in kilometers instead of millimeters.

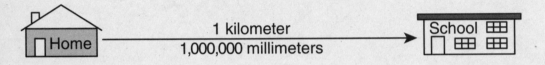

Think about it. If you lived 1 kilometer from school, then that would mean you lived 1,000,000 millimeters from school. Instead of writing all those zeros, it is easier and simpler to say 1 kilometer, so that's the unit you should use.

The same rule applies to other types of measurements. Imagine that you have a giant pile of rocks and you want to find out how much the pile of rocks weighs. Should you use ounces or tons to describe the weight of the rocks? 1 ounce weighs about the same as 5 quarters.

1 ton = 32,000 ounces

If the pile of rocks weighed 1 ton, it would be the same as 32,000 ounces. Describing the pile of rocks as "1 ton" is simpler than saying "32,000 ounces," so you should use tons as the unit of measurement.

Know It All! Elementary School Math

Estimating Measurements

You can use everyday objects around you to help you **estimate** (guess) measurements. Here are some examples.

The height of a classroom doorway is about 2 meters or $6\frac{1}{2}$ feet.

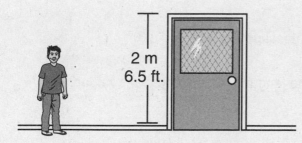

2 m
6.5 ft.

The length of your forearm from your knuckles to your elbow is about 1 foot or 30.5 centimeters.

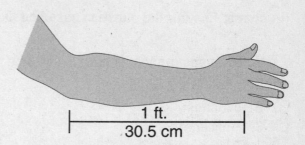

1 ft.
30.5 cm

The distance from the tip of your thumbnail to the first joint on your thumb is about 1 inch or 2.5 centimeters.

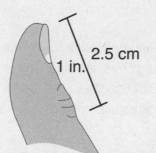

2.5 cm
1 in.

The average textbook weighs about 2 pounds or 1 kilogram.

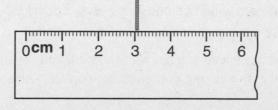

2 lb.   1 kg

A dime is about 1 millimeter thick.

0 cm 1 2 3 4 5 6

Zip It!

Whitcomb L. Judson first came up with the idea for a "hookless fastener" when a friend was having trouble tying his shoes. Judson designed a product that would allow people to fasten their shoes with one hand. His product was patented in 1893. Today, we call Judson's invention the zipper.

Directions: Answer the question based on the passage Zip It!

▶ Which measurement is most likely to be the length of a common jeans zipper?

- ○ A 15 centimeters
- ○ B 15 inches
- ○ C 15 millimeters
- ○ D 15 kilometers

Know It All Approach

Read the question carefully. Picture what the zipper on your jeans looks like. Read answer choice (A). Could a jeans zipper be 15 centimeters long? Yes, it could. There are about 30 centimeters in 1 foot, so 15 centimeters would be about 6 inches long. Keep this answer choice, but read the rest of the answer choices to make sure this answer choice is the best one.

Look at answer choice (B). Could a jeans zipper be 15 inches long? No! That's more than a foot long. Get rid of this answer. Look at answer choice (C). Could a jeans zipper be 15 millimeters long? No! 1 millimeter is about the thickness of 1 dime. 15 millimeters is only 1.5 centimeters, which would be a *very* tiny zipper. Get rid of answer choice (C). Look at answer choice (D). Could a jeans zipper be 15 kilometers? No way! A kilometer is more than half a mile. 15 kilometers would be one *loooooong* zipper! Get rid of answer choice (D).

Now that you have ruled out the wrong answer choices, you can be sure that answer choice (A) is correct. Fill in the bubble that goes with answer choice (A).

Directions: Read the passages below and answer the questions that follow.

Coming Right Up!

Stevie Starr has an unusual talent. He can swallow items whole and then spit them back up in whatever order he chooses. In the past, Starr has swallowed and spit up billiard balls, house keys, lightbulbs, live goldfish, rings, and many other items. Would you want to wear a ring that had been in some stranger's stomach? That's nasty!

1. Which unit of measurement would you use to describe the weight of a ring?
   - ○ A liter
   - ○ B gram
   - ○ C pound
   - ○ D meter

2. Which of the following could describe the length of a house key?
   - ○ A 6 centimeters
   - ○ B 6 inches
   - ○ C 6 meters
   - ○ D 6 millimeters

Bring In the Noise

For years, the popular theater group STOMP has used its dance moves and rhythmic beats to entertain audiences all over the world. The cast members of STOMP use everyday items like trashcans, brooms, pens, and even matchbooks to beat out toe-tapping rhythms onstage as they dance. Their rhythms can be as quiet as rustling papers or almost loud enough to bring down the roof. Either way, STOMP has been a crowd-pleaser since 1991.

3. Which of the following would best help you estimate the length of a broom handle?
   - ○ A a paper clip
   - ○ B a rock
   - ○ C a forearm
   - ○ D a dime

Subject Review

In this chapter, you learned how to choose the best unit to describe measurements. You also learned how to estimate measurements. Review what you've learned.

When you measure an item, make sure you use the best unit of measurement for the job. Here are some hints.

- Use kilometers or miles to measure long distances, and use centimeters or inches to measure the length of small objects.
- Use units like liters, pints, and quarts to measure liquid volume.
- Use tons to describe the weight of huge objects, and use grams, ounces, or pounds to describe the weight of small to medium-size objects.

You can use everyday objects around you to help you **estimate** (guess) measurements. Here are some examples.

- The height of a classroom doorway is about 2 meters or 6.5 feet.
- The distance from the tip of your thumbnail to the first joint on your thumb is about 1 inch or 2.5 centimeters.
- The length of your forearm from your knuckles to your elbow is about 1 foot or 30.5 centimeters.
- The average textbook weighs about 2 pounds or 1 kilogram.
- A dime is about 1 millimeter thick.

Here are the answer to the questions on page 125.

Who first came up with the idea for the zipper?
Whitcomb L. Judson designed and patented the first zipper in 1893.

What bizarre talent does Stevie Starr have?
He can swallow common items whole and then spit them back up in a specific order.

What theater group uses garbage cans and brooms in its performances?
The cast members of the show STOMP use everyday objects to beat out rhythms on the stage.

CHAPTER 17
Graphs and Data

What object was voted the favorite invention of all time?

Are falling stars really stars?

When did author R.L. Stine begin writing?

In this chapter, you'll learn about different types of graphs. You'll also learn how to read and use these graphs.

Graphs and Data

Data (or information) are often shown in the form of a diagram. Diagrams that show data are called **graphs.** There are many different types of graphs. The examples that follow show the different ways a group of friends can record data about their bug-collecting activities.

A **bar graph** uses bars of different lengths to show data. Here is a bar graph.

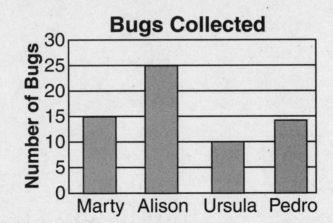

The title of the graph tells you what the graph is about—collecting bugs. The *x*-axis shows who collected the bugs. The *y*-axis tells you how many bugs were collected. The bars on the graph show that Marty collected 15 bugs, Alison collected 25 bugs, Ursula collected 10 bugs, and Pedro collected 14 bugs. Notice that Pedro's bar doesn't fall directly on a line. You need to make a rough guess in his case.

A **line graph** uses lines to show how data have changed over a period of time.

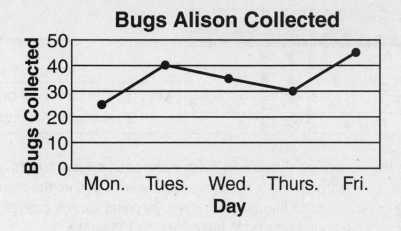

Bugs Alison Collected

The line graph above shows how many bugs Alison collected each day Monday through Friday. Notice how the line graph shows you Alison's progress over time. For example, she collected only 25 bugs on Monday, but collected 45 bugs on Friday.

A **circle graph** is a circle that is divided to show parts of a whole. The entire circle represents the whole, and each section inside it stands for a part of the whole.

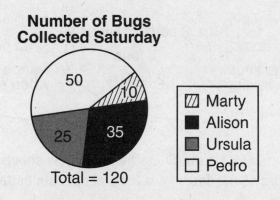

Number of Bugs Collected Saturday

This circle graph shows how many of the 120 total bugs each person collected. Pedro collected 50 of the 120 bugs. Marty collected only 10 of the 120 bugs. Notice the box to the right of the graph. This is the **key.** The key shows you what each color or pattern on the graph represents. In this case, each color stands for a different person.

A **pictograph** uses symbols to show data.

Types of Bugs the Group Collected Saturday	
Spiders	🕷 🕷
Grasshoppers	🕷 🕷 🕷
Butterflies	🕷 🕷 🕷 🕷 🕨
Moths	🕷 🕷 🕨

Key: 🕷 stands for 10 bugs.
🕨 stands for 5 bugs.

The key beside the pictograph shows how many items each symbol stands for. A full bug stands for 10 units. A half-bug stands for 5 units. You need to add up the units in the graph to find the total number of each bug type collected. By doing so, you can tell that the group collected 20 spiders, 30 grasshoppers, 45 butterflies, and 25 moths.

A **Venn diagram** uses overlapping circles to organize groups of objects. This diagram organizes facts about butterflies and moths.

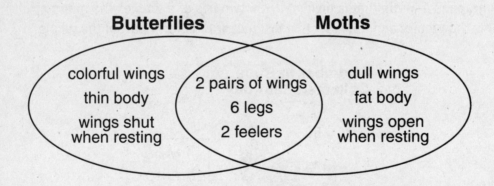

Butterflies   **Moths**

colorful wings
thin body
wings shut when resting

2 pairs of wings
6 legs
2 feelers

dull wings
fat body
wings open when resting

The left circle shows facts about butterflies. The right circle shows facts about moths. The center part where the circles overlap shows facts about both butterflies and moths.

Know It All! Elementary School Math

You Can't Ride a Computer to School

In 2002, a British radio station asked its listeners to vote for their all-time favorite invention. You'll never guess which invention got 70% of the vote. The winner was (drum roll, please) . . . not the computer . . . not the lightbulb . . . but the *bicycle*!

Directions: Answer the question based on the passage You Can't Ride a Computer to School.

▶ The fourth grade-teachers asked their students to name their favorite inventions. The graph below shows the results of the survey.

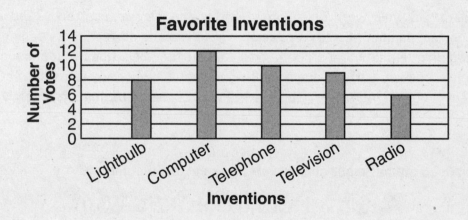

What is the total number of students who voted in the survey?

Answer: _____ students

Know It All Approach

Read the question and the graph carefully. The title and labels on the graph will tell you what the graph is showing. The *x*-axis tells you what inventions the students voted for. The *y*-axis tells you how many votes each invention received.

To figure out the *total* number of students who voted in the survey, read each bar and figure out the number of students each bar stands for. The first bar shows that 8 students voted for the lightbulb. The second bar shows that 12 students voted for the computer. The third bar shows that 10 students voted for the telephone. The fourth bar falls between lines 8 and 10. This means that 9 students voted for the television. The fifth bar shows that 6 students voted for the radio. To find the total number of students who voted, add all of those numbers together. 8 + 12 + 10 + 9 + 6 = 45.

Now, double-check your work. Did you read the bars on the graph correctly? Did you write down the correct number of votes for each bar? Did you add correctly? Try adding the numbers together in a different order than you did the first time. Does your second answer match the first one you got? If it does, you can be sure that your answer is correct. A total of 45 students voted in the survey. Finally, write your answer, "45 students," neatly on the line.

Directions: Read the passages and answer the questions that follow.

Catch a Falling Star? No Way!

 Did you know that falling stars aren't really stars at all? Stars are big balls of gas in space. The things we call falling stars are actually solid chunks of space matter. When these chunks of matter enter Earth's atmosphere, they burn up. That's what causes the trail of light we see from the ground.

1. Georgia and her parents spent a few days camping in the mountains. At night, Georgia kept track of how many falling stars she saw. The chart below shows the number of falling stars Georgia saw each night she camped.

Falling Stars Seen on Camping Trip

Night	Number
Thursday	III
Friday	ℍℍ II
Saturday	IIII
Sunday	II

Using the information from the table, make a bar graph on the grid below to show how many stars Georgia saw each night of her camping trip.

Be sure to

- give the graph a title
- label the *x*-axis and *y*-axis
- graph all the data from the chart
- use an appropriate scale

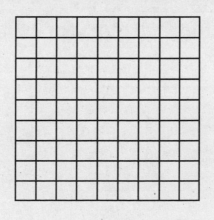

Scary Stories

Popular writer R.L. Stine has written more than 100 books for children and young adults. He first began writing when he was just 9 years old. As a child, he wrote short stories and gave them to his classmates to read. If you're thinking about becoming a writer, it's never too soon to start!

2. Jeremy, Toshi, Suzi, and Wendy are big fans of R.L. Stine novels. They made a circle graph to show how many R. L. Stine books they had each read. Which bar graph shows the same information as the circle graph?

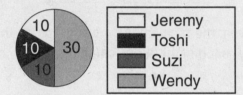

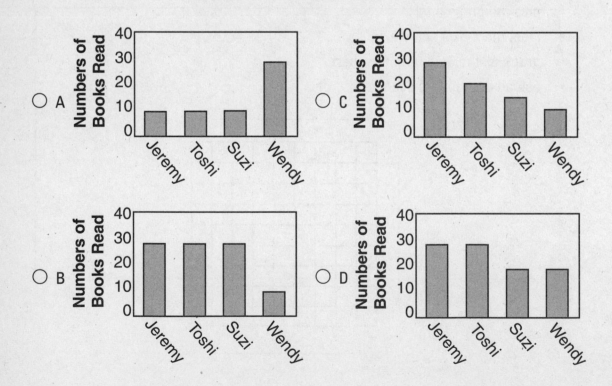

3. Suzi wants to become a writer like R.L. Stine when she is older. She spends time each day writing in her journal. The table below shows how much time Suzi spends writing each day.

Time Suzi Spent Writing

Day	Time (hours)
Monday	0.5
Tuesday	1
Wednesday	0.5
Thursday	1
Friday	2

In the space below, draw a pictograph that shows the amount of time Suzi spends writing. Make sure you include a title and a key.

Subject Review

Graphs are diagrams that show data.

There are many types of graphs.

- A **bar graph** uses bars of different lengths to show data.
- A **line graph** uses lines to show how data have changed over a period of time.
- A **circle graph** is a circle that is divided to show parts of a whole. The entire circle represents the whole, and each section inside it stands for a part of the whole.
- A **pictograph** uses symbols to show data.
- A **Venn diagram** uses overlapping circles to organize groups of objects.

Some graphs have a **key.** The key shows you what each color or symbol in the graph represents.

It is very important to read and understand all parts of a graph. Read the title, the *x*-axis, the *y*-axis, the key, and any other information the graph includes.

Here are the answers to the questions from the beginning of the chapter.

What object was voted the favorite invention of all time?

In a 2002 British radio survey, 70% of the listeners voted for the bicycle as their all-time favorite invention.

Are falling stars really stars?

Falling stars are not stars. Real stars are balls of gas. Falling stars are chunks of solid matter that burn up when they hit Earth's atmosphere.

When did author R.L. Stine begin writing?

He began writing when he was 9 years old. He used to write stories for his class-mates.

CHAPTER 18
Probability

Which actor gets hit on the head more often than a carpenter's nail?

What were the first crayons made of?

How did people open cans before can openers were invented?

If you flip a coin, what are the chances of the coin coming up heads? In this chapter, you will learn how to make simple predictions like this.

Predicting Simple Events

Probability is a prediction you make that measures the chances of an event occurring.

Think about the coin mentioned in the introduction. If you flip a coin, what are the chances of the coin falling with the heads side facing up?

Here's how a fraction can show this probability.

$$\text{Number of favorable (good) outcomes possible} \longrightarrow \frac{1}{2} \longleftarrow \text{Number of total outcomes possible (good and bad)}$$

The top number of the fraction is 1 because there is only 1 heads side of a coin. The bottom number is 2 because there are 2 sides of a coin in all—heads and tails.

You can write the same probability as a percent, too. Just convert the fraction $\frac{1}{2}$ to a percent like you normally would. $\frac{1}{2}$ is equal to 50%, so the probability of the coin coming up heads is 50%.

Probability can also be shown as a **ratio.** A ratio is similar to a fraction, but it's written a different way. Think of it as a fraction turned sideways.

$$\frac{1}{2} \quad = \quad 1:2$$
$$\text{Fraction} \qquad \text{Ratio}$$

The ratio 1:2 is read as "one to two." The number of good outcomes is on the left side of the colon. The total number of possible outcomes (good and bad) is on the right side of the colon. The numerator of the fraction will always come first if it is written as a ratio.

Know It All! Elementary School Math

Using Words to Describe Probability

You can use the words below to describe probability.

- certain
- more likely
- equally likely
- less likely
- impossible

Suppose that you have a bag full of different alphabet letters. The list below shows the contents of the bag.

7 *A*s

3 *B*s

1 *C*

3 *D*s

5 *E*s

Now imagine that you reached into the bag—no peeking!—and pulled out one letter. Which word from the list would best describe the chances of your pulling out the letter *A*? If you said *more likely*, you'd be correct. There are more *A*s in the bag than any other letter.

Which word from the list would best describe the probability of your pulling out the letter *Z*? Well, that would be *impossible*, because there are no *Z*s in the bag.

You'd be *equally likely* to pull out a *B* as you would a *D*, because there are 3 *B*s and 3 *D*s in the bag.

You would be *less likely* to grab a *B* than you would an *E*, because there are 5 *E*s and only 3 *B*s in the bag.

If the bag had nothing in it but *A*s, you would be *certain* to pull out an *A*.

Hong Kong Headache

Few people are busier than actor and martial artist Jackie Chan. Over the past 40 years, Jackie Chan has appeared in more than 90 films. Even more amazing is that he does all of his own stunts in his movies. He has fallen from buildings several stories tall, dangled from cranes, and been hit on the head more times than anyone can count. He has joked that he has broken every bone in his body. Of course, breaking bones is no laughing matter. The only one who should attempt Jackie Chan's death-defying stunts is Jackie Chan himself!

Directions: Answer the question based on the passage Hong Kong Headache.

▶ A video rental store is giving away free movie posters of martial arts stars. The posters are all rolled up individually in a box. The box holds 10 Jet Li posters, 7 Jackie Chan posters, and 12 Bruce Lee posters.

What is the probability of choosing a Jackie Chan poster without looking?

○ A $\frac{7}{22}$

○ B $\frac{1}{7}$

○ C $\frac{7}{29}$

○ D $\frac{22}{29}$

Know It All Approach

Read the question carefully and look for important words and numbers you need in order to figure out the answer. In this question the important information is *10 Jet Li posters, 7 Jackie Chan posters, 12 Bruce Lee posters,* and *probability.*

To answer this question, you first need to add all of the posters together. This tells you the total number of outcomes in the box. 10 + 7 + 12 = 29. So 29 is the bottom number of your probability fraction. You have to figure out the chances of pulling a Jackie Chan poster from the box. According to the question, there are 7 Jackie Chan posters in the box. 7 is the number of possible good outcomes, so it is the top number of your probability fraction. The correct answer is $\frac{7}{29}$. Double-check your answer to make sure the numbers you have used match the numbers in the question. Make sure you have added correctly. When you have ruled out all of the other answer choices, fill in the bubble that goes with answer choice (C).

Directions: Read the passages below and answer the questions that follow.

Color It Oily

 When crayons were first invented, they were very different from the neat, waxy ones used today. Crayons were originally made in Europe more than 100 years ago. In their early days, they were made of a mixture of charcoal and oil. Imagine how messy that must have been!

1. Hector is coloring a design. He is pulling crayons randomly out of a box without looking. The box has 4 black crayons, 2 red crayons, 2 blue crayons, 1 yellow crayon, and 1 green crayon.

Part A

Write a fraction that describes Hector's chances of pulling a black crayon out of the box.

Answer: _____

Part B

Use one of the following words or phrases to describe Hector's chances of pulling a yellow crayon out of the box: *certain, more likely, less likely, impossible*

Answer: _____

Chisel Before Serving

 Did you know that people used tin cans to keep food fresh before there were can openers? The tin can (like soup cans today) was invented in 1810. The can opener wasn't invented until around 1858. Before then, people had to use a hammer and chisel to open their cans!

2. A box has 4 cans of soup, 8 cans of spinach, 7 cans of peas, and 3 cans of corn. What is the probability of choosing a can of soup without looking into the box?
    - ○ A 1:4
    - ○ B 4:18
    - ○ C 1:22
    - ○ D 4:22

Subject Review

Probability is a prediction you make that measures the chances of an event occurring.

You can show probability as a fraction, a percent, or a ratio.

A **ratio** is similar to a fraction, but it's written differently. For example, the probability fraction $\frac{1}{2}$ is written 1:2 as a ratio.

You can also use words to describe probability.

- certain
- more likely
- equally likely
- less likely
- impossible

Remember the questions on page 141? Here are the answers.

Which actor gets hit on the head more often than a carpenter's nail?

Jackie Chan, movie actor and martial artist, does all of his own dangerous action stunts and often gets injured.

What were the first crayons made of?

The first crayons were made of charcoal and oil. Imagine having to wash your hands after every drawing!

How did people open cans before can openers were invented?

Before the can opener was around, people had to use a hammer and chisel to open cans.

Directions: Read the passages below and answer the questions that follow.

Crazy Claws!

 The therizinosaurus, or "scythe lizard," was a very unusual beast. This 3-ton dinosaur had huge claws that measured about 28 inches long. One scythe lizard fossil found in 1961 even had claws that were 36 inches long! If you had lived in dinosaur times, you wouldn't have had to worry about those claws, though. Scientists are pretty sure that the scythe lizard was not a meat eater. Instead, they believe that the scythe lizard used its long claws to strip plants of their leaves.

1. Dinosaur experts measured some therizinosaurus claws. They recorded their findings in the chart below.

Dinosaur Claw Length

Claw Label	Claw Length
Claw A	$1\frac{1}{4}$ feet
Claw B	38 inches
Claw C	1 yard
Claw D	32 inches

> 12 inches = 1 foot
> 3 feet = 1 yard

Which of the claws measured was the longest?

- ○ A Claw A
- ○ B Claw B
- ○ C Claw C
- ○ D Claw D

2. Which of the following metric units would the scientists have used to measure the length of the therizinosaurus claws?

- ○ A kilometers
- ○ B liters
- ○ C grams
- ○ D centimeters

3. A team of dinosaur experts spent a month digging up fossils. The graph below shows their progress.

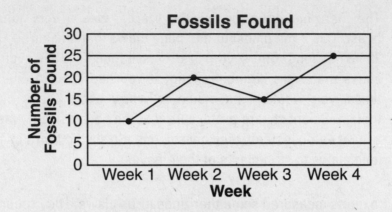

Part A

How many fossils did the scientists find altogether?

Answer: _____ fossils

Part B

In which week did the scientists find more fossils than in Week 1 but fewer fossils than in Week 2?

Answer: _____

4. A scientist had a bag full of fossil teeth. The bag contained 7 canines, 20 molars, and 4 incisors. Which of the following words or phrases describes the scientist's chances of pulling a molar out of the bag without looking?

○ A certain
○ B more likely
○ C less likely
○ D impossible

A New Twist on Sporting Events

Some New Zealand sports fans are trying to organize an event called the Fringe Games. The Fringe Games will be an international competition like the Olympics, but with very different sporting events. Athletes will compete in sports like sideways running, gymnastics races, and a type of bicycle ballet.

5. Ian wanted to organize his own extreme sports competition. He asked kids in his neighborhood to sign up for the events they wanted to compete in. The chart below shows the events that Ian's friends signed up for.

Sporting Events

Event	Number of Participants
Sideways running	ⅢⅢ
Bicycle ballet	|||
Gymnastics races	||||
Unicycle race	ⅢⅢ|||

Using the information from the chart, make a bar graph on the grid below to show how many kids signed up for each competition. Be sure to

- give the graph a title
- label the *x*-axis and *y*-axis
- graph all the data
- use an appropriate scale

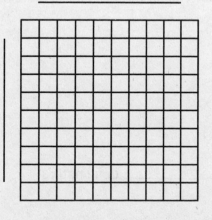

6. Use your ruler to help you solve this problem. Ian drew a diagram of the area in which the athletes would compete, as shown below.

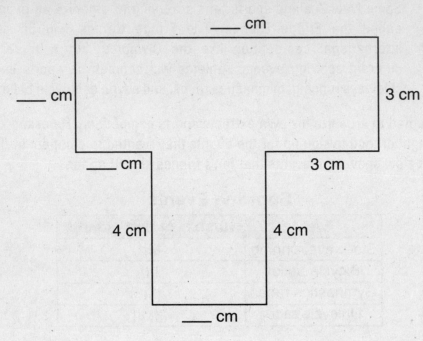

_____ cm

_____ cm

3 cm

_____ cm

3 cm

4 cm

4 cm

_____ cm

Part A

Measure the sides of Ian's diagram. Write the missing measurements on the lines provided above.

Part B

What is the perimeter of Ian's diagram?

Show your work.

Answer: _____ centimeters

Answer Key for Chapters and Brain Boosters

Chapter 2

1. 1865
2. C
3. D
4. B

Chapter 3

1. 4 playing stones
2. 30 crewmembers
   Top-scoring answers should describe each step taken to answer this question. Your description should make sense to whoever reads it, and not just you. Don't skip any steps! Here is a sample answer.
   The question asks how many crewmembers would be evenly divided on each ship. First add the numbers of crewmembers together: 24 + 26 + 40 = 90. There are 90 crewmembers altogether. Now divide 90 by the number of ships: 90 ÷ 3 = 30. There would be 30 crewmembers on each ship.

Chapter 4

1. A
2. D
3. commutative property

Chapter 5

1. D
2. A

Brain Booster 1: Review of Chapters 2–5

1. Part A: Nathan
   Part B: Vera
2. C
3. A
4. D
5. Top-scoring answers should clearly explain how you can tell whether the number sentence is correct without doing the math. Here is a sample answer.
   This number sentence is correct. You can tell without doing the math because of the commutative property. In multiplication problems, it doesn't matter what order you multiply the numbers in. In this problem, the same numbers are multiplied together on both sides of the equal sign; they're just in a different order.
6. 24,950; 25,463; 25,649; 26,942
7. C

Chapter 6

1. B
2. C

Chapter 7

1. Michael
2. B
3. D

Chapter 8

1. 45
2. D
3. A

Chapter 9

1. D
2. A
3. B
4. C

Brain Booster 2: Review of Chapters 6–9

1. C
2. 50%
3. A
4. Part A: Manuela
   Part B: Justin
5. To get a top score for your response, give the correct answer and explain how you arrived at that answer. Here is a sample answer.
   It will take Teddy 3 more climbs to reach 14 feet. I got this answer because the problem says Teddy got 2 feet higher every time he climbed, so I just counted by 2s. If he climbed 8 feet on his last try, the next time he would climb 10 feet, the time after that he would climb 12 feet, and the time after that he would climb 14 feet. That's 3 more tries.
6. D
7. A

Chapter 10

1. B
2. D

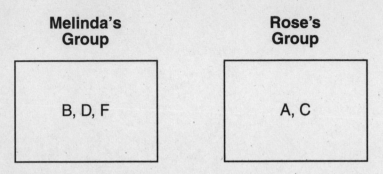

Melinda's Group

B, D, F

Rose's Group

A, C

3.

To get a top-scoring answer, you must explain why you grouped the shapes in Melinda's group together and why you grouped the shapes in Rose's group together. Explain it in a way that makes sense to the test-grader, not just yourself. Here are some sample answers.

Melinda's Group: *The shapes in this group go together because they are all squares and rectangles. They all have right angles and two pairs of parallel sides.*

Rose's Group: *The shapes in this group go together because they are all trapezoids. These shapes all have only one pair of parallel sides.*

Chapter 11

1. B
2. D
3. cylinder

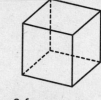

4.
   6 faces
   12 edges

Chapter 12

1. (2,2)
2. D

Chapter 13

1. A
2. B

Chapter 14

1. 50 inches
2. D

Brain Booster 3: Review of Chapters 10–14

1. A
2. D
3. Make sure you draw your rectangular pyramid neatly. You should be able to see through it so you can count all of the faces and edges.

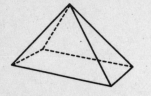

   5 faces
   8 edges
4. the mall
   (3,5)
5. C
6. 168 inches

Chapter 15

1.

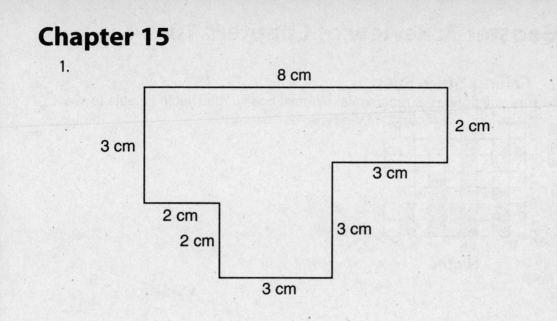

2. B

Chapter 16

1. B
2. A
3. C

Chapter 17

1.

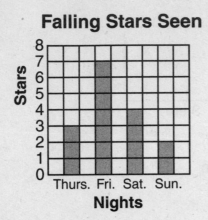

Falling Stars Seen

2. A

3.

Amount of Time Suzi Spent Writing

Monday	🕐
Tuesday	🕐
Wednesday	🕐
Thursday	🕐
Friday	🕐 🕐

Key: 1 🕐 = 1 hour

Chapter 18

1. Part A: $\frac{4}{10}$

   Part B: less likely

2. D

Brain Booster 4: Review of Chapters 15–18

1. B
2. D
3. Part A: 70 fossils

   Part B: Week 3
4. B
5.

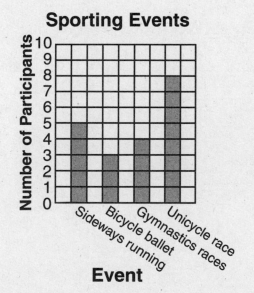

6. Part A:

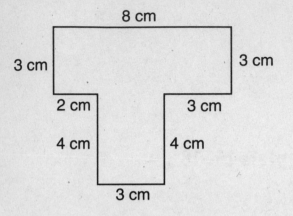

Part B: Remember to show all of your work neatly! The answer is 30 centimeters. See the work shown below.

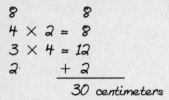

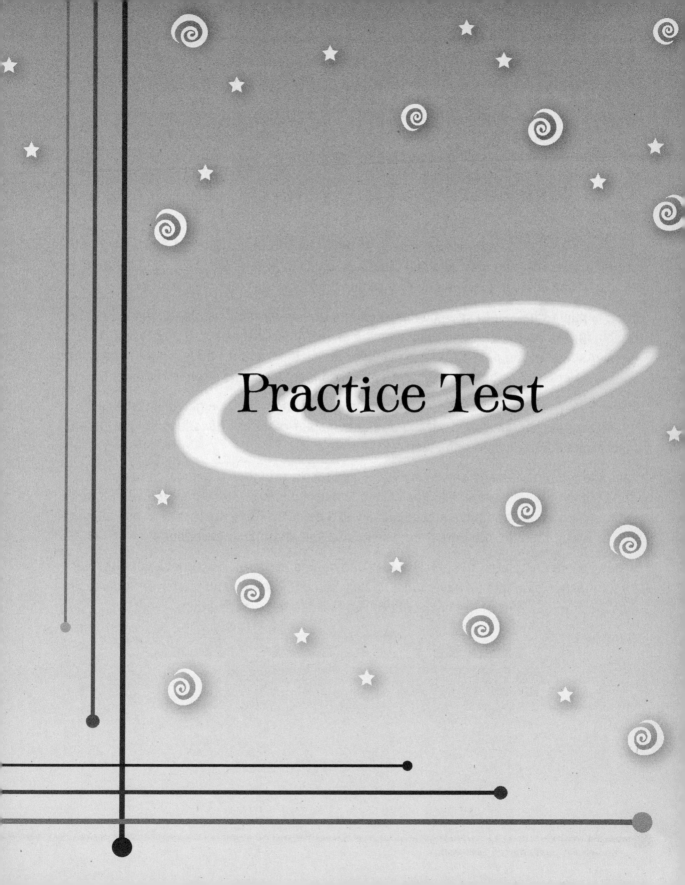

Practice Test

Introduction to the *Know It All!* Practice Test

By now you've reviewed all the important skills that you should know for elementary school math. You know how to solve word problems (chapter 3). You know how to identify two-dimensional figures such as rectangles and triangles (chapter 10). You also know how to read information on a graph and answer questions about it (chapter 17). These are just a few examples that don't even include all the great pieces of information you've picked up along the way. You *Know It All!*

If you're ready, it's time to try out the skills from the eighteen chapters in this book in a practice test. This test may be similar to a test you take in class. It contains multiple-choice, short-answer, and open-response questions.

Each multiple-choice question on the test has four answer choices. You should fill in the bubble for the correct answer choice on the separate answer sheet. Cut or tear out the answer sheet on the next page, and use it for the multiple-choice questions. You can write your answers to the short-answer and open-response questions directly in the test.

The practice test contains forty-two questions, including thirty-two multiple-choice questions, five short-answer questions, and five open-response questions. Give yourself ninety minutes to complete the test.

Take the practice test the same way you would take a real test. Don't watch television, don't talk on the telephone, and don't listen to music while you take the test. Sit at a desk with a few pencils, and have an adult time you if possible. Take the test all in one sitting. If you break up the test into parts, you won't get a real test-taking experience.

When you've completed the practice test, you can go to page 189 to check your answers. Each question also has an explanation to help you understand how to answer it correctly. Don't look at this part of the book until you've finished the practice test!

Good luck!

Answer Sheet

1. (A) (B) (C) (D)
2. (A) (B) (C) (D)
3. (A) (B) (C) (D)
4. Use space provided.
5. (A) (B) (C) (D)
6. Use space provided.
7. (A) (B) (C) (D)
8. (A) (B) (C) (D)
9. (A) (B) (C) (D)
10. Use space provided.
11. (A) (B) (C) (D)
12. Use space provided.
13. (A) (B) (C) (D)
14. (A) (B) (C) (D)
15. (A) (B) (C) (D)
16. Use space provided.
17. (A) (B) (C) (D)
18. (A) (B) (C) (D)
19. (A) (B) (C) (D)
20. (A) (B) (C) (D)
21. (A) (B) (C) (D)

22. Use space provided.
23. (A) (B) (C) (D)
24. (A) (B) (C) (D)
25. (A) (B) (C) (D)
26. (A) (B) (C) (D)
27. (A) (B) (C) (D)
28. Use space provided.
29. (A) (B) (C) (D)
30. (A) (B) (C) (D)
31. (A) (B) (C) (D)
32. (A) (B) (C) (D)
33. (A) (B) (C) (D)
34. Use space provided.
35. (A) (B) (C) (D)
36. (A) (B) (C) (D)
37. (A) (B) (C) (D)
38. (A) (B) (C) (D)
39. (A) (B) (C) (D)
40. Use space provided.
41. Use space provided.
42. (A) (B) (C) (D)

Directions: Use the following passage to answer questions 1–4:

Don't Touch That!

Did you know that there is a glass museum? No, it isn't made out of glass! Rather, the Corning Museum of Glass displays glass items. The museum's collection includes pieces from all over the world, including some that are almost 3,500 years old.

The museum includes pieces of all sizes and colors. All in all, it houses more than 35,000 glass items. You can look at—but not touch—the fragile items in this museum.

1. According to the passage, the Corning Museum of Glass has more than 35,000 glass items. Which answer is the same as 35,000?

   A   thirty-five
   B   three hundred fifty
   C   three thousand, five hundred
   D   thirty-five thousand

2. Some items at the Corning Museum of Glass are cylinder-shaped. Which drawing shows a cylinder?

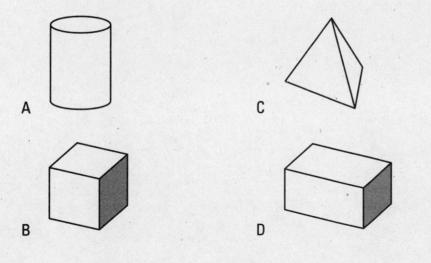

A

C

B

D

3. Bobby's favorite drinking glass has a pattern design around the rim. The pattern is shown below. Choose the figure that would come next in the pattern.

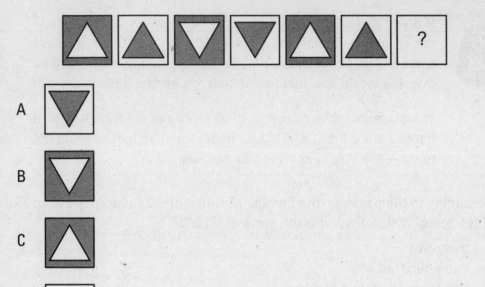

A

B

C

D

Know It All! Elementary School Math

4. The Corning Museum of Glass has items created in the following years:

   1876   1662   880   1970   150

   Write the years in order, from least to greatest.

5. Janet drew two figures. One showed a rectangle inside a hexagon. The other showed a triangle inside a trapezoid. Which answer could be Janet's drawings?

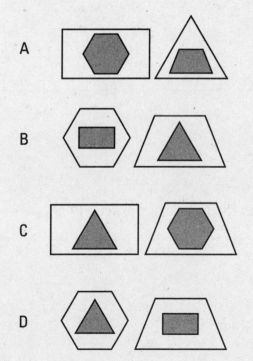

A

B

C

D

6. Juan went to a produce stand to buy cantaloupes. He saw this sign at the market.

Part A

Juan wants to buy 6 cantaloupes. Which will cost him less, buying 6 separate cantaloupes or buying 1 box of 6 cantaloupes?

Show your work.

Answer: _____

Part B

What would be the **least** expensive way for Juan to buy exactly 10 cantaloupes?

Show your work.

Answer: _____

Directions: Use the following passage to answer questions 7–10:

More Bamboo, Please!

 A full-grown panda is $5\frac{1}{4}$ feet to 6 feet long and weighs up to 330 pounds. It's hard work for pandas to maintain their weight. That's because they mostly eat bamboo, which is very low in nutrients. In order to get enough calories and vitamins to live, a panda must eat more than 20 to 40 pounds of bamboo each day. It takes the panda 10 to 16 hours to eat that much bamboo!

7. At the age of 1, a panda weighs 75 pounds. How much do three 1-year-old pandas weigh?

   A   25 pounds
   B   75 pounds
   C   150 pounds
   D   225 pounds

8. A school group of 100 people visited the San Diego Zoo. 10 people in the group were teachers; the other 90 were students. What percent of the people in the group were students?

   A   0.9%
   B   9%
   C   90%
   D   900%

9. At birth, a baby panda weighs $\frac{1}{4}$ pound. Which fraction is **greater** than $\frac{1}{4}$?

   A $\frac{1}{2}$

   B $\frac{1}{6}$

   C $\frac{1}{8}$

   D $\frac{1}{10}$

10. Write a number sentence that will help you find out how much bamboo a panda must eat per hour in order to eat 30 pounds of bamboo in 12 hours.

    _____

11. Ms. Thompson shows her class two shapes, a square and a regular triangle. The length of each side of the square is 15 inches. The perimeter of the triangle equals the perimeter of the square. If the three sides of the triangle are equal in length, what is the length of one side of the triangle?

A   5 inches
B   10 inches
C   20 inches
D   60 inches

12. This is a two-part question.

Part A

A Chihuahua is one of the world's smallest dogs. In fact, a full-grown Chihuahua weighs only 6 pounds. How many **ounces** does a full-grown Chihuahua weigh? (1 pound = 16 ounces)

Show your work.

Answer: _____

Part B

The Great Dane is one of the world's largest dogs. A full-grown male Great Dane is about 32 inches high at the shoulder. How many **feet** tall is a full-grown male Great Dane at the shoulder? (1 foot = 12 inches)

Show your work.

Answer: _____

Directions: Use the following passage to answer questions 13–16:

Delicious and Good for You, Too!

Joker

Looking for a snack that's delicious and good for you? Try a mango! These red and green tropical fruits are juicy, tangy, and sweet. They're great chopped up in a salsa or simply cut into slices. They also make a tasty addition to any salad.

Mangoes are full of nutrients that help your body grow strong. The graph below shows the daily nutritional requirements filled by just one mango. Because it is high in sugar, a mango is an excellent source of energy. So have a mango today. Just make sure to have an adult help you cut it into bite-size pieces!

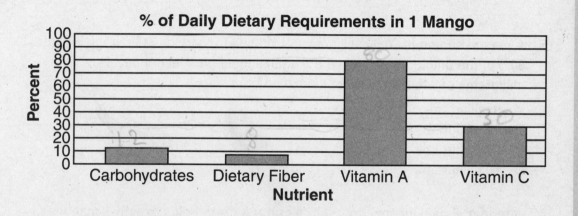

13. A mango provides exactly 30 percent of which daily dietary requirement?

A   carbohydrates
B   dietary fiber
C   vitamin A
D   vitamin C

14. Which table contains the same information that is shown in the graph on the previous page?

A

Nutrient	Percent
Carbohydrates	30
Dietary Fiber	80
Vitamin A	8
Vitamin C	12

C

Nutrient	Percent
Carbohydrates	12
Dietary Fiber	80
Vitamin A	30
Vitamin C	8

B

Nutrient	Percent
Carbohydrates	8
Dietary Fiber	12
Vitamin A	30
Vitamin C	80

D

Nutrient	Percent
Carbohydrates	12
Dietary Fiber	8
Vitamin A	80
Vitamin C	30

15. A person should have 60 milligrams of vitamin C every day. A mango provides 30% of the amount of vitamin C a person should have every day. How many milligrams of vitamin C are in one mango?

A   9 milligrams
B   18 milligrams
C   50 milligrams
D   60 milligrams

16. A farmer is packing mangoes into a crate. A crate holds exactly 8 mangoes. The farmer has 210 mangoes to pack. How many crates will the farmer need in order to pack all of the mangoes?

Show your work.

Answer: _____

17. Robert Wadlow was the tallest person who ever lived. Which is most likely Robert Wadlow's height?

   A   2.72 millimeters
   B   2.72 centimeters
   C   2.72 meters
   D   2.72 kilometers

18. The map below shows a portion of Silo City.

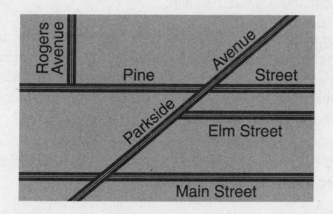

   Which two streets on the map are **not** parallel to each other?

   A   Pine Street and Rogers Avenue
   B   Pine Street and Elm Street
   C   Main Street and Elm Street
   D   Main Street and Pine Street

Spaced Out

In one sense, the planets Jupiter and Mars are next-door neighbors. After all, Mars is the fourth planet from the sun and Jupiter is fifth. You couldn't really say that they're close neighbors, though. At their closest, Jupiter is 335 *million* miles from Mars. That's more than two times the distance between Mars and the Sun. Most of the time the two planets are much, much farther apart than that.

In the night sky, Mars is a rich red color, while Jupiter is beautifully striped in blue, pink, red, and brown. From Earth they appear about the same size, but that's because Mars is so much closer. In reality, Jupiter is more than 10,000 times larger than Mars!

19. According to the passage, Jupiter and Mars are 335 million miles apart. Which is the correct way to write 335 million?

    A 3,350
    B 335,000
    C 3,350,000
    D 335,000,000

20. Jupiter has 8 times as many moons as Mars. Jupiter has 16 moons. Use the equation below to solve for m, the number of moons Mars has.

$$8m = 16$$

    A 1
    B 2
    C 4
    D 8

21. Because Mars's gravity is different from Earth's, a person's weight on the two planets is not the same. A person's weight on Mars is roughly $\frac{2}{5}$ of his or her weight on Earth. Shipra weighs 80 pounds on Earth. What would she weigh on Mars?

   A   20 pounds
   B   25 pounds
   C   32 pounds
   D   40 pounds

22. Two NASA Mars Exploration Rovers are expected to land on Mars's surface in January 2004. While on Mars, each rover will travel 100 meters a day as it searches the Martian surface. How far will the two rovers travel **together** in 3 days?

Show your work.

Answer: _____

23. Jerry solved the multiplication problem below as follows:

$$28 \times 9 = 252$$

Which expression could Jerry use to find out whether he solved the problem correctly?

A $252 \div 9$

B 252×9

C $252 + 9$

D $252 - 9$

24. Which of the following digits has a line of symmetry?

A 0

B 2

C 4

D 9

Directions: Use the following passage to answer questions 25–28:

Take a Dive

Have you ever wondered why some swimming pools are shaped like the letter L? The reason that many parks and clubs build L-shaped pools has to do with swimmer safety. With an L-shaped pool, one part of the pool can be roped off for diving only. A separate diving area is the safest way for divers and swimmers to share a pool.

The diving area is always the deepest part of a swimming pool. The water in diving areas is at least 8 feet deep. Usually the water is much deeper. The deep water allows you to dive off the board safely, without scraping the bottom of the pool. Diving is a fun way to cool off and enjoy the swimming pool. Just make sure to follow all the rules so that your dive will be safe as well as fun!

Know It All! Elementary School Math

25. The story says that a diving area should be at least 8 feet deep. How many **inches** are equal to 8 feet?

A 8 inches

B 48 inches

C 80 inches

D 96 inches

26. Sharon went swimming for $10\frac{1}{2}$ minutes. Which decimal is the same as $10\frac{1}{2}$?

A 10.14

B 10.25

C 10.4

D 10.5

27. Danielle has an inflatable pool toy. When inflated, the pool toy has 6 faces. Which of the following could be Danielle's pool toy?

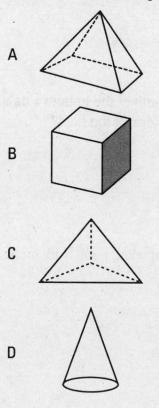

A

B

C

D

28. Roberto swims in an L-shaped pool at a local park. The diagram below shows some of the dimensions of the pool.

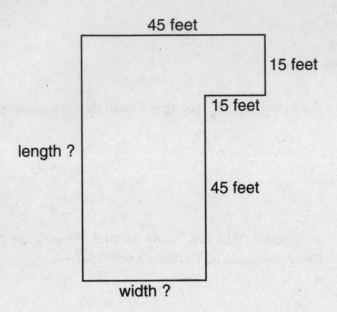

45 feet

15 feet

15 feet

length ?

45 feet

width ?

Part A

In the diagram, the length of the left side and the width of the bottom side are not given. What are the length and width of these sides of the pool?

Length = _____ feet

Width = _____ feet

Part B

Roberto likes to walk along the perimeter of the swimming pool. How many feet long is the perimeter of the pool?

Show your work.

Answer: _____ feet

29. A tapir is a large land animal that looks a little like a pig and a little like an anteater. Which unit of measurement could you use to find the mass of a tapir?

A kilograms
B kiloliters
C kilometers
D milliliters

30. Three of the expressions below are equal. Which one is **different** from the other three?

A 2×4
B $2 + 2 + 2 + 2$
C $2 \times 2 \times 2 \times 2$
D $4 + 4$

Say It with Sunflowers

If you have a hankering to do some gardening, perhaps you should consider growing sunflowers. Sunflowers are easy to grow, so they're good for beginning gardeners. Even better, they are quite striking when they are fully grown. The flower of a sunflower plant is a gorgeous bright yellow. You might need a ladder to see the flower, though; some sunflowers can grow up to 12 feet tall!

Sunflowers should be planted in the spring. Anytime after the threat of overnight frost has passed is fine. Plant your seeds one to two inches deep in soil. Don't plant your sunflowers too close together. They have deep root systems and need room to grow. You should see your flowers poking through the soil a week or two after you plant them. Four months later, you'll have gorgeous, full-grown sunflowers to enjoy.

31. Andrew planted 20 sunflower seeds. 16 of the seeds grew into sunflowers; the other 4 did not. What percent of the seeds Andrew planted grew into sunflowers?

   A   16%
   B   36%
   C   80%
   D   90%

32. Barrett's sunflower is growing 3 inches a day. She realizes she can calculate how many days it will take for the sunflower to grow 24 inches using the expression

$$3x = 24$$

where x is the number of days. How many days will it take for Barrett's sunflower to grow 24 inches?

   A   3
   B   8
   C   12
   D   21

33. Joan planted four sunflowers. The table below shows the height of each sunflower when it was full-grown.

Sunflower A	3.45 meters
Sunflower B	3.50 meters
Sunflower C	3.98 meters
Sunflower D	3.36 meters

Which sunflower was the **second** tallest?

A   Sunflower A
B   Sunflower B
C   Sunflower C
D   Sunflower D

34. Michael decides to plant sunflowers in his backyard. He plants the sunflower seeds 2 feet apart. He plans to plant the flowers in a row that is 10 feet long.

The diagram below shows how Michael plants the seeds. The number line represents the 10-foot row in which Michael plants the seeds. The two *x*s represent the first two seeds Michael planted.

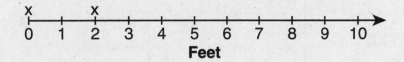

How many flowers will Michael plant in his 10-foot row?

Answer: _____

35. A number cube has the numbers 1, 2, 3, 4, 5, and 6 on its faces. What is the probability that an even number will land face up when the number cube is rolled?

A $\frac{1}{6}$

B $\frac{2}{6}$

C $\frac{3}{6}$

D $\frac{4}{3}$

36. What are the coordinates of the star in the coordinate grid below?

A (2,9)

B (5,4)

C (6,5)

D (8,2)

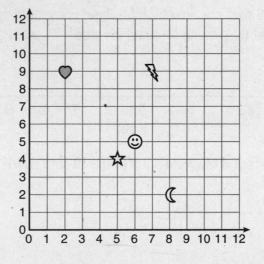

Directions: Use the following passage to answer questions 37–40:

There's a Fungus Among Us

Every year, the people of Crystal Falls, Michigan, have a one-of-a-kind celebration. It's called the Humongous Fungus Festival. What, you may ask, is a humongous fungus? It's a gigantic mushroom. The one growing in Crystal Falls is thought to be the world's largest. The massive mushroom covers 38 acres of land. It weighs about 100 *tons!* (No one knows for sure because there isn't a scale anywhere in the world big enough to weigh it!) It is believed to be between 1,500 and 10,000 years old.

The humongous fungus grows beneath a forest. Most of it lives underground, so it is not visible to tourists. Some visitors are disappointed, because they expect to see a giant mushroom towering in the sky. The people of Crystal Falls aren't disappointed, though. Every year they celebrate the fungus with three days of contests, games, feasts, and a big fireworks display. The people of Crystal Falls are happy to report, "There's a fungus among us!"

37. The humongous fungus weighs about 100 tons. Which of the following most likely weighs about the same as the humongous fungus?

A   the planet Earth
B   a textbook
C   an apple
D   a whale

38. Tonya measured the height of four plants she found in the forest where the humongous fungus grows. The table shows the heights of the plants she measured.

Plant A	$1\frac{1}{3}$ yards
Plant B	50 inches
Plant C	3.5 feet
Plant D	4 feet 1 inch

Which plant was the tallest?

A   Plant A
B   Plant B
C   Plant C
D   Plant D

39. Armando wants to enter the horseshoes tournament at the Humongous Fungus Festival. It costs $3.00 to enter the contest. He has $1.65. How much more money does Armando need in order to enter the contest?

    A $0.35
    B $0.45
    C $1.35
    D $1.45

40. Clarice's hometown had a festival. At the festival were 48 rides. Clarice decided she wanted to ride every ride at the festival. She also decided that she wanted to ride an equal number of rides every day she attended the festival.

Part A

If Clarice attends the festival for 2 days, how many rides must she ride each day?

Show your work.

Answer: _____ rides

Part B

If Clarice attends the festival for 3 days, how many rides must she ride each day?

Show your work.

Answer: _____ rides

41. The game of Evens is played with a spinner. If a spin lands on an even number, the player wins. If it lands on an odd number, the player loses.

Bobby is playing Evens. He has his choice of any of the three spinners below.

Which spinner gives Bobby the best chance of winning?

Answer: _____

Explain why the spinner you chose gives Bobby the best chance of winning the game of Evens.

_____

_____

_____

_____

42. Look at the four shapes below.

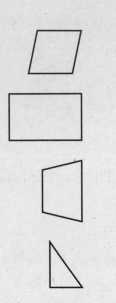

Which shape would look exactly the same after it was flipped over the dotted line?

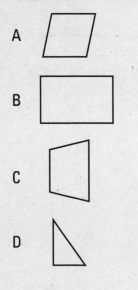

A

B

C

D

Answers and Explanations for the Practice Test

Answers and Explanations for the Practice Test

1. **D** This question asks you to write out a number in word form. To write out the number 35,000, count the places in the number. It has 5 places, which means that 35,000 goes to the ten thousands place. This number is written as thirty-five thousand, answer choice (D).

2. **A** Identify the cylinder. Remember, a cylinder has two circular bases. A soup can is an example of a cylinder. Answer choice (A) shows a cylinder. Answer choice (B) shows a cube. Answer choice (C) shows a triangular pyramid. Answer choice (D) shows a rectangular prism.

3. **B** To answer this question, you must figure out the pattern. Each figure in the pattern is a square with a triangle inside. One way to figure out a pattern is to look at each part of the pattern separately. The squares are shaded, white, shaded, white, shaded, and white. The next figure must be shaded. Look at the answer choices; you can get rid of (A) and (D) because they are white squares. Next, look at the direction of the triangles. The pattern is up, up, down, down, up, and up. The next shape must be down. That leaves answer choice (B) as the best answer choice.

4. **150, 880, 1662, 1876, 1970**
   To put numbers in order, first make a list of the numbers. Make sure to line up all the digits in the same place value. In your list, only two numbers have 3 digits. Those numbers are smaller than the numbers with 4 digits, so compare them first. 150 has a 1 in the hundreds column; 880 has an 8 in the hundreds column. 1 is smaller than 8, so 150 is smaller than 880. The first two numbers are 150 and 880. Continue this way to put the final three numbers in order.

5. **B** This question asks you to find two shapes. Don't look for both shapes at the same time. Look for one shape first, and get rid of any answer choice that does not show that shape. Let's look at the answer choices. Which answer choices have a rectangle inside a hexagon? Only answer choice (B), so (B) must be the correct answer. If you had trouble naming the different shapes, go back and review the names of shapes in chapter 10.

6. Read the information on the sign and use it to answer the questions. Make sure to show your work, or you will not receive full credit on the question. Here is an example of an answer that earns the most points possible.
Part A
A box of 6 cantaloupes costs $9.50.
6 separate cantaloupes cost 6 × $1.75 = $10.50.
It would cost Juan less to buy 1 box of 6 cantaloupes.
Part B
Juan should buy 1 box of six cantaloupes for $9.50 and 4 cantaloupes for $1.75 each. This will cost him a total of $9.50 + (4 × $1.75) = $9.50 + $7.00 = $16.50. Juan could also buy 10 cantaloupes by paying for them separately. If he buys 10 cantaloupes separately, he would pay 10 × $1.75 = $17.50, which would be more expensive.

7. **D** If one young panda weighs 75 pounds, then three young pandas should weigh 3 times as much. Multiply 75 by 3. 3 × 75 = 225. Three young pandas weigh 225 pounds.

8. **C** The problem tells you that 90 out of 100 people who visited the zoo were students. 90 out of 100 can be written as the fraction $\frac{90}{100}$, which is the same as 90%. That means that 90% of the people who visited the zoo were students.

9. **A** To answer this question, you will have to compare fractions. One way to compare fractions is to draw a shape and divide it into fourths. Then shade in $\frac{1}{4}$ to represent the weight of the baby panda. Next, draw the same figure, divide it into halves, and shade in $\frac{1}{2}$. Answer choice (A) looks like it is greater than $\frac{1}{4}$. To be sure, draw figures for each of the other answer choices and compare it to $\frac{1}{4}$. You should find that $\frac{1}{4}$ is greater than answer choices (B), (C), and (D), so (A) is correct.

10. **30 ÷ 12 = ?** The words *per hour* give you a clue that your number sentence should include division. To find out how much bamboo a panda would eat in one hour in order to eat 30 pounds in 12 hours, you must divide 30 by 12.

11. **C**   This question includes a lot of information. The best way to answer it is to look at each piece of information carefully to figure out what it tells you. Then, move on to the next piece of information. The problem says that the length of each side of the square is 15 inches. You know the perimeter of the square must be the sum of all four sides. Because all four sides of a square have the same length, the perimeter of this square is 15 × 4 = 60 inches.

The perimeter of the square and the perimeter of the triangle are the same, so the triangle also has a perimeter of 60 inches. Each side of the triangle has the same length, so the length of one side is 60 ÷ 3 = 20 inches.

12.   You need to convert pounds to ounces and inches to feet to answer this question. If you have trouble with this question, go back and review measurements in chapter 15. Here is an example of an answer that may receive the most possible points.
Part A
1 pound = 16 ounces
6 pounds = 6 × 16 = 96 ounces
A full-grown Chihuahua weighs 96 ounces.
Part B
1 foot = 12 inches

$32 \text{ inches} = 32 \div 12 = 2\frac{8}{12} \text{ feet}$

$2\frac{8}{12} \text{ feet} = 2\frac{2}{3} \text{ feet}$

A full-grown Great Dane is $2\frac{2}{3}$ feet tall.

13. **D**   Which bar shows exactly 30 percent of the daily dietary requirement? To find the answer, put your finger on the 30 on the y-axis. Then run your finger across the graph until you find the bar that stops exactly at that line. Don't be fooled by bars that stop below your finger or by the bar that goes past your finger. The one that stops exactly at 30% is vitamin C.

14. **D**   Answers from one question may help you with another question. In question 13, you learned that a mango provides exactly 30 percent of the vitamin C you need in one day. Which table agrees with what you just learned? Only the table for answer choice (D) agrees with what you found for question 13. All of the other tables have different numbers for vitamin C. Because you know that the correct answer must say that vitamin C = 30%, you know that (D) must be the correct answer.

15. **B**   This question tells you that a person needs 60 milligrams of vitamin C every day. It also tells you that a mango provides 30 percent of the vitamin C a person needs. To calculate the amount of vitamin C in one mango, you need to calculate 30 percent of 60. $0.3 \times 60 = 18$. Be careful with the decimal point! The correct answer is 18 milligrams.

16. **The farmer will need 27 crates to pack the mangoes.**   This question asks you to divide 210 mangoes into crates that contain 8 mangoes each. Solve the problem using division, and watch out for those sneaky leftover mangoes!
$210 \div 8 = 26$ r2
26 of the crates will be completely full. The farmer will need the 27th crate for the 2 leftover mangoes.

17. **C**   This question asks you to use your knowledge of measurements to guess the height of Robert Wadlow. It asks about metric measurements, which may or may not be as familiar to you as U.S. customary measurements. Millimeters and centimeters are very small units. No human being has ever been 2.72 millimeters or 2.72 centimeters tall, and certainly not the tallest person who ever lived! Kilometers are very large units of measure. A kilometer is a little more than a half a mile. No human has ever been 2.72 kilometers tall—that would be more than 1 mile! The best answer choice must be (C). (By the way, Robert Wadlow was 8 feet 11 inches tall.)

18. **A**   This question asks you to look at a map and find the two streets that are not parallel to each other. Lines that are parallel will never touch. Find the pairs of streets that are listed in each answer choice and see if they are parallel. Answer choice (A) lists two streets that touch each other, so they cannot be parallel. This looks like the correct answer. Find the streets for the other answer choices and check that they are parallel, and thus not the correct answer.

19. **D**   This question asks you to find the number 335 million among the answer choices. Review your place values. The first three digits take you up to the hundreds. The next three digits take you up to the hundred thousands. The millions start at the seventh digit and continue all the way through the hundred millions in the ninth digit. For the number 335 million, you will need a nine-digit number that starts with 335. The correct answer is 335,000,000.

20. **B** Find the number you can substitute for *m* to make 8*m* = 16 true. Remember that 8*m* is a way of writing 8 × *m*. What number can you multiply by 8 to get 16? If you said 2, you were right. If you didn't know the answer, don't give up. Substitute each of the answer choices for *m* and choose the one that makes the number sentence true. For example, answer choice (A) is 1. Is 8 × 1 equal to 16? No, so (A) cannot be the correct answer. Check all of the answer choices that way. If you get rid of one or two answer choices, great! You'll have a good chance of guessing the right answer. If you get rid of three, even better—the only answer choice left will be the correct answer.

21. **C** The question tells you that on Mars, Shipra would weigh $\frac{2}{5}$ what she weighs on Earth. It also tells you that she weighs 80 pounds on Earth. To find how much she would weigh on Mars, figure out what $\frac{2}{5}$ of 80 is. You might want to convert $\frac{2}{5}$ to a decimal before you calculate. $\frac{2}{5} \times 80 = 0.4 \times 80 = 32$. Shipra would weigh 32 pounds on Mars.

22. **600 meters** You know that each rover travels 100 meters per day. There are 2 rovers and they travel for 3 days.
100 × 2 = 200 meters, the distance the two rovers travel together in 1 day
200 × 3 = 600 meters, the distance the two rovers travel together in 3 days

23. **A** This question asks you to remember how you can check your work on a multiplication problem. To check a multiplication problem, divide the product by one of the factors. If you've forgotten how to check your work on math problems, go back and review chapter 4.

24. **A** A line of symmetry is a line that can divide a figure exactly in half so that each half is a mirror image of the other. 0 has two lines of symmetry—it can be divided exactly in half either vertically or horizontally. None of the other numbers shown in the problem has a line of symmetry. That's why (A) is the correct answer to this question.

25. **D** To answer this question, you must convert feet to inches. First, remember how many inches are in a foot. If you said 12 inches equal 1 foot, you were correct! Now, if there are 12 inches in 1 foot, how many inches are in 8 feet? 12 × 8 = 96. The correct answer is 96 inches.

26. **D** You may remember that $\frac{1}{2}$ is equal to 0.5. Another way to convert the fraction to a decimal is to find an equivalent fraction with a denominator of 10. $\frac{1}{2} = \frac{5}{10}$. Another way to write $\frac{5}{10}$ is 0.5. Don't forget the whole-number part. The answer is 10.5, answer choice (D).

27. **B** Count the faces of each drawing in the answer choices. Get rid of answers that do not have 6 faces. Answer choice (A) is a rectangular pyramid, which has 5 faces. Get rid of that answer choice. Answer choice (B) is a cube; a cube has 6 faces, so keep this answer. Answer choice (C) is a triangular pyramid, which has 4 faces. Get rid of (C), too. Answer choice (D) is a cone. Cones have only 1 face. Get rid of (D). The correct answer must be (B).

28. To answer this question, you must figure out the missing length and width of the pool. You must also figure out the perimeter of the pool.

    Put your finger on the part of the drawing marked "length?" and look across to the right side of the drawing. You should see two lengths marked vertically on the drawing. One says "15 feet" and the other says "45 feet." If you add these two lengths together, they will be exactly equal to the unknown length of the pool. The length of the pool is 60 feet.

    Now put your finger on the part of the drawing marked "width?" and look up at the horizontal side of the pool marked "45 feet." Look at the smaller horizontal part of the pool marked "15 feet." You should see that if you subtract the smaller width from the larger one, what's left over is the unknown width of the pool. The width of the pool is 30 feet.

    Here is an example of an answer that received the most possible points.

    Part A
    Length = 60 feet
    Width = 30 feet
    Part B
    The perimeter of the pool is 45 + 15 + 15 + 45 + 30 + 60 = 210 feet.

29. **A** To answer this question, you must recognize a metric unit of mass. All of the other answer choices are not units of mass. A kiloliter is used to measure the volume of liquids. Kilometers and millimeters are used to measure length. Only a kilogram is used to measure mass, so (A) must be the correct answer choice.

30. **C** Read each of the answer choices to see which expression is not equal to the other answer choices. Answer choice (A) is 2 × 4, which is equal to 8. Answer choice (B) is 2 + 2 + 2 + 2, which is also equal to 8. Answer choice (C) is 2 × 2 × 2 × 2, which is equal to 16. Answer choice (D) is 4 + 4, which is equal to 8. The only answer choice that is not equal to 8 is (C).

31. **C** 16 of Andrew's 20 sunflower seeds grew into sunflowers. To find out what percent of Andrew's sunflower seeds grew into sunflowers, write a fraction that shows this information: $\frac{16}{20}$. Then convert the fraction to a percent. $\frac{16}{20} = 0.80 = 80\%$. The correct answer is 80%, answer choice (C).

32. **B** Find the number that you can substitute for x to make $3x = 24$ true. Remember that $3x$ is a way of writing $3 \cdot x$. What number do you multiply by 3 to get 24? If you said 8, good for you! If you weren't sure, you could test all the answers. Does 3 × 3 equal 24? No, so answer choice (A) cannot be the correct answer. Test all of the answer choices this way and get rid of the ones you know are incorrect.

33. **B** Study the numbers in the chart. First, find the tallest sunflower. Sunflower C is the tallest at 3.98 meters. But wait—the question asks you to identify the *second* tallest sunflower. Which sunflower is shorter than 3.98, but taller than the other sunflowers? If you said sunflower B, you're correct! Sunflower B is 3.50 meters tall, taller than 3.45 and 3.36.

34. **6 flowers** You must count by twos to figure out where Michael will plant his other seeds. The question and the number line show that Michael is planting his seeds 2 feet apart. The number line shows that Michael planted his last seed at the 2-foot mark. So just count by twos and put an x over the 4, 6, 8, and 10 on the number line. Now count the total number of marks above the number line. There are marks over the 0, 2, 4, 6, 8, and 10. How many marks is that? If you said 6, give yourself a pat on the back!

35. **C**   The even numbers on the cube are 2, 4, and 6. Remember probabilities from chapter 18: The total number of good outcomes is the top number of the fraction. There are 3 even numbers on the number cube, so the top number of your fraction is 3. The bottom number of a probabilities fraction is the total number of possible outcomes (good and bad). Your bottom number should be 6, because there are 6 sides on the number cube. So the correct answer choice is (C), $\frac{3}{6}$.

36. **B**   Find the star on the graph. Use your finger to follow the line the star sits on down to the number on the *x*-axis. The star is even with the number 5 on the *x*-axis, so 5 is the first number in the coordinate pair. Now use your finger to follow the line the star sits on over to the number on the *y*-axis. The star is even with the number 4 on the *y*-axis, so 4 is the second number in your coordinate pair. Answer choice (B), (5,4), is the correct answer.

37. **D**   Think back to the conversion charts in chapter 15. You may remember that 1 ton is the same as 2,000 pounds. The humongous fungus weighs about 100 tons, which would be the same as 200,000 pounds! Look at answer choice (A). Do you think the planet Earth weighs 200,000 pounds? No, it must weigh more than that if it has a fungus weighing that much sitting on it. Get rid of (A). What about answer choice (B), a textbook? No, a textbook weighs about 2 pounds. Get rid of (B). What about answer choice (C), an apple? No, an apple weighs less than a pound. Get rid of (C). What about (D), a whale? Well, a whale is an enormous animal that certainly weighs much more than an apple or a textbook. It also weighs much less than the planet Earth. Answer (D), a whale, is the answer that makes the most sense.

38. **B**   To answer this question, convert all of the measurements in the chart to feet. Plant A is $1\frac{1}{3}$ yards, so that's 4 feet. Plant B is 50 inches, which is equal to 4 feet 2 inches. Plant C is 3.5 feet. Plant D is 4 feet 1 inch. Now that you have converted all of the plant heights to feet, which one is the tallest? If you said plant B, at 4 feet 2 inches, you're correct! Good work!

39. **C**   This is a subtraction question. To figure out how much more money Armando needs, subtract $1.65 (the amount of money he has) from $3.00 (the total amount of money he needs). $3.00 − $1.65 = $1.35. So answer choice (C), $1.35, is correct.

40. To answer this question, you must divide the number of rides by the number of days Clarice will attend the festival. To answer part A, you must divide 48 rides by 2 days. How many rides will Clarice ride each of the 2 days? If you said 24, you're correct. Now answer part B. To figure out how many rides Clarice will ride in 3 days, divide 48 by 3. $48 \div 3 = 16$.

Part A **24 rides**

Part B **16 rides**

41. To answer this question, you must look closely at the spinners. The question asks you to identify the spinner that gives Bobby the best chance of landing on an even number.

Look at spinner A. It is divided into 4 equal parts, 2 even and 2 odd. The even numbers take up $\frac{1}{2}$ of the spinner.

Look at spinner B. 2 is the only even number on the spinner, and it takes up only $\frac{1}{3}$ of the spinner.

Look at spinner C. 2 and 4 take up more than $\frac{1}{2}$ of the spinner. So which spinner gives Bobby the best chance of landing on an even number? If you said spinner C, you're correct. Why? Because the even-numbered sections take up more area on the spinner than the odd-numbered sections.

Here's an explanation that may receive the maximum points for this question.

Spinner C would give Bobby the best chance to win Evens because it has the greatest area for even numbers. It is more likely for the spinner to land on an even number than to land on an odd number.

42. **B** To answer this question, you must remember that a flip creates a mirror image on the other side of the line. Look at shape (A). Will it look the same when flipped? No, it will be slanted to the left when it is flipped instead of being slanted to the right. Look at shape (B). Will it look the same when flipped? Yes! Because it is made of all right angles, it will look exactly the same when flipped. Keep this answer choice, but look at the other ones to be on the safe side. Look at shape (C). Will it look the same when flipped? No, the longest vertical line will be on the left when flipped instead of on the right. Look at shape (D). Will it look the same when flipped? No. The right angle will be on the right-hand side instead of on the left. So answer choice (B) is correct!

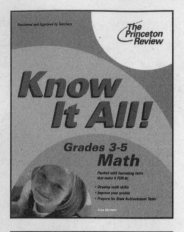